Culture and the Cognitive Science of Religion

Culture and the Cognitive Science of Religion is the first book to bring together cultural psychology and the cognitive science of religion (CSR). Containing much-needed discussion of how good research should do more than simply follow methodological prescriptions, this thought-provoking and original book outlines the ways in which CSR can be used to study everyday religious belief without sacrificing psychological science.

Cresswell's pragmatist approach expands CSR in a radically new direction. The author shows how language and culture can be integrated within CSR in order to achieve an alternative ontogenetic and phylogenetic approach to cognition and argues that a view of cognition that is not based on modularity, but on the dynamic connection between an organism and its milieu, can lead to a view of evolution that makes much more room for the constitutive role of culture in cognition.

As a provocative attempt to persuade researchers to engage with religious communities more directly, the book should be essential reading for academics, researchers and postgraduate students, as well as psychologists interested in the cognitive science of religion, theological anthropology, religious studies and cultural anthropology.

James (Jim) Cresswell is an associate professor and program chair in psychology at Ambrose University (Calgary, Canada).

Cultural Dynamics of Social Representation

The series **Cultural Dynamics of Social Representation** is dedicated to bringing the scholarly reader new ways of representing human lives in the contemporary social sciences. It is a part of a new direction – cultural psychology – that has emerged at the intersection of developmental, dynamic and social psychologies, anthropology, education, and sociology. It aims to provide cutting-edge examinations of global social processes, which for every country are becoming increasingly multi-cultural; the world is becoming one 'global village', with the corresponding need to know how different parts of that 'village' function. Therefore, social sciences need new ways of considering how to study human lives in their globalizing contexts. The focus of this series is the social representation of people, communities, and – last but not least – the social sciences themselves.

For more information about this series, please visit www.routledge.com.

In this series

Optimizing the Self
Social Representations of Self-help
Ole Jacob Madsen

Imagination in Human and Cultural Development
Tania Zittoun and Alex Gillespie

Representing Development
Pasts, Presents and Futures of Transformative Models
Edited by David Carré, Jaan Valsiner and Stefan Hampl

Persons and their Minds
Towards an Integrative Theory of the Mediated Mind
Svend Brinkmann

Culture and the Cognitive Science of Religion
James Cresswell

Culture and the Cognitive Science of Religion

James Cresswell

Routledge
Taylor & Francis Group
LONDON AND NEW YORK

First published 2018 by Routledge

2 Park Square, Milton Park, Abingdon, Oxfordshire OX14 4RN
52 Vanderbilt Avenue, New York, NY 10017

Routledge is an imprint of the Taylor & Francis Group, an informa business

First issued in paperback 2019

British Library Cataloguing-in-Publication Data
A catalogue record for this book is available from the British Library

Library of Congress Cataloguing-in-Publication Data
A catalog record for this book has been requested

ISBN: 978-1-138-21939-7 (hbk)
ISBN: 978-0-367-36338-3 (pbk)

Typeset in Times New Roman
by Apex CoVantage, LLC

Thanks to my mother, Linda Bliss.

Contents

Series editor's preface

Making sense of unseen orders

> Were one asked to characterize the life of religion in the broadest and most general terms possible, one might say that it consists of the belief that there is an unseen order, and that our supreme good lies in harmoniously adjusting ourselves thereto.
>
> (James, 1902, p. 53)

Some relevant topics have escaped their deserved research focus in our contemporary psychology. Religion is one of those. It is one of the most sophisticated inventions of the human species as it involves features that for any science are difficult to grasp. The notion of "unseen order" is conceptually a hard task for any science – not only for psychology. The unexpected outcomes in the unraveling of the overwhelming "junkness" of the genetic code or the solid strength of the graphene layer show how sciences discover unexpected features in such unseen orders. Hence, there is much to see for psychology in its looking at phenomena of religious sentiments – another domain where the origins of order is unseen but the consequences of that order govern the lives and deaths of very real human beings.

Sometimes it is hard to see the unseen at very close distance. The affect-laden nature of human existential beliefs – all of which are of central importance in religions – makes research efforts on them too close for comfort. How can I rationally analyze the basic irrationality of my existence as a person? Every effort in that direction is tampered by my own desire to modify my own ways of existing. I am – so to say – *self-inclusive*. Psychology in general has had to encounter that issue in many of its approaches – and psychology of religious experiences is one of those.

Psychology has been a science of paradoxes (Valsiner, 2012) – one of which is the recognition of the religious sentiments as central for its empirical efforts. Historically, our contemporary oversight of religious experience has not been the case in the history of psychology. Just the contrary – up to the 1920s, it was a major theme for thinking and curiosity by psychologists. Almost every psychologist (obviously excluding those who adhered to their deeply fixed belief in behaviorism) took interest in the complex phenomena of religious feelings and thinking. A century later, however, phenomena of religion are being systematically investigated only by few psychologists, and only from very few perspectives. Theoretical fashions – rather than theoretical creativity – dominate the field. This is more than a show of basic intellectual poverty of the discipline as the social world in which we all are embedded is burgeoning with complex issues anchored in a variety of religious convictions. The examples of French schoolgirls' self-decoration acts by wearing headscarves (Bowen, 2007) that became interpreted as religious self-declarations on the one hand, and the atrocities committed by the "Islamic State" against human beings and cultural heritages, on the other – are all evidence of the absolute centrality of religious sentiments in human lives and societies. And psychology as a science got its own "head start" five centuries ago when Martin Luther revolted against the neoliberal trend in the dominant Catholicism to raise funds for various earthly projects by selling insurance policies against sinfulness in the form of indulgencies. Such business practices irritated the deep believers and led to the Reformation in Europe. Aside from promoting literacy and setting the stage for further emergence of capitalism, the resulting impact on human minds was profound – the search of religiosity in one's own mind. Such a turn toward inner reflexivity about matters of the faith led to theorizing the complexities of the human soul – and to the birth of "soul science" of psychology in the eighteenth and nineteenth centuries. No surprise that issues of psychology of religious experiences occupied the interests of nineteenth-century psychologists – with William James in the lead (James, 1902).

The present book is a refreshing new look at religion – through the lens of a cultural psychology of ecological kind – that of *enactivism* (Baerveldt & Verhaggen, 2012). This perspective within others in the new area of cultural psychology is particularly potent for turning the otherwise-static area of social representations into a dynamic perspective. Building on the dynamic systems theory, the social presentations used in religious belief systems become understood as functional tools for human self-reflection. This has not been the case in cognitive studies of religion that have been focused on the centrality of categories to which the religious experiences could be reduced. Cognitive psychology's efforts to represent religions through systems of fixed categories have led to overlook of the dynamic, generative roles that inventions of religious sentiments play in societies and individual minds. At relevant moments of existential stress, new deities are created – and followed. This sets up interesting feed-forward loops that operate in non-religious practices as well (Valsiner, 1999). *We* set up our deities, invent rituals that – for us at least – prove that they exist – and then claim that we follow *their* orders. My *will to believe* leads to my *believing in* – whichever deity or any other unseeable order I might elect to be the object of my believing-in.

Religious practices are enacted – they are dramatizations created by the persons for their own purposes. Rituals are the key framework within which the connection between external acting and intra-psychological self-organization is being established. As such, they are constantly re-constructed – albeit in a relatively stable way. Rituals emerge in human play (Köpping, 1998) and are standardized in communal co-action as normative systems. A look at social *re*presentations that support rituals begins from the social *pre*sentations that constitute parts of the emerging new rituals. A developmental perspective on the emergence of deep and pervasive beliefs is the necessary foundation for any cultural psychology of religion. The present book makes that imperative very clear.

Cultural psychologies of today explicitly reject the axiom of genetic determinacy of higher mental actions that is at the root of the contemporary fashion for the glory of neurosciences. Without denying the neurological basis for higher psychological functions – among which religious phenomena constitute the most complex domain – the explanation of the complexity of human subjectivity by its reduction to brain mechanisms is unwarranted. Different levels of organization have their own ways of regulating their dynamics – the brain functions in ways of its own, and the psyche operates by organizational forms that are not reducible to those of the brain. Each level of complexity *builds upon* its nearest neighbors but also adds new features. The human psyche is *generative* – constantly adding new varieties of human experience to the previous ones. This happens through mediation by signs – the target field of the science of semiotics. Religious phenomena require analyses in semiotic rather than neuroscientific terms.

The major interlocutor for the author of this book is the field of cognitive science of religion. In the latter, the assumptions of the biological basis of the "cognitive architectures" prevail. In contrast to that assumption, the focus on the centrality of fluency in the sign systems of the given community is set forth. Such fluency is worked out in ontogeny – religious thinking takes place in the context of participation in the affective practices of the community. And such participation is characterized by dynamic flow of personal experiences that are constantly leading to ever-new versions of forms of belief. Some of these remain transitory, but others become fixed. Once they do become strongly held opinions, a religious belief system becomes socially dangerous. When opinions dominate, thinking ends – and the road to radicalization of the given religious belief system becomes open. Any religious system can be turned into a dogmatic basis for social stigmatization and annihilation under the circumstances of radicalization. The problem of religiously acclaimed inter-human conflicts that proliferate in our worlds today is not in the religion as such, but the fixation of its ideas and radicalization of its followers who consider the fixed ideas as legitimizing any kinds of actions.

Jaan Valsiner
Berlin, June 2017

References

Baerveldt, C., & Verhaggen, T. (2012). Enactivism. In J. Valsiner (Ed.), *The Oxford handbook of culture and psychology*. New York: Oxford University Press.

Bowen, J. (2007). *Why the French do not like headscarves*. Princeton, NJ: Princeton University Press.

James, W. (1902). *The varieties of religious experience*. New York: Longmans Green and Co.

Köpping, K-P. (Ed.). (1998). *The games of Gods and man: Essays in play and performance*. Hamburg: LIT Verlag.

Valsiner, J. (1999). I create you to control me: A glimpse into basic processes of semiotic mediation. *Human Development*, *42*, 26–30.

Valsiner, J. (2012). *The guided science: History of psychology in the mirror of its making*. New Brunswick, NJ: Transaction Publishers.

Preface

Billig's (1996) *Arguing and Thinking* introduced me to a unique way of looking at psychology – which I will get to shortly – but it had a comment in the preface that has stuck with me for quite some time. He advocated taking time to wander through the shelves in a library and to read different books that one found. Billig was extolling the virtues of being surprised and discovering ideas that can be missed when one only works within the discourse of a particular field within a particular discipline. The importance of setting oneself up to be surprised is a message that lingers with me. The way that my project herein began was surprising and something that I had no intention of pursuing. Theoretical psychology and its close cousins – critical and cultural psychology – are the fields that drive my interest. One colleague expressed a desire for me to be a "meat and potatoes" psychologist, and by this he meant that I ought to be a traditional mainstream empiricist researcher. Another colleague came to my office and handed me an advertisement for a seminar at Calvin College on the cognitive science of religion and Christianity, which looked like "meat and potatoes" psychology. I sent in an application and found myself at a seminar with cognitive psychologists, and I am thankful to the seminar leader, Justin Barrett, because the whole experience set me up to be surprised.

The surprise came from the fact that I was sitting at a seminar with several mainstream psychological researchers. I hinted above how I take an approach to psychology that concerns itself with culture in a very pervasive way. I had been reading Billig because he argues that social actions are what make up psychological functioning. Psychological phenomena are topics about which people debate in life and coordinate together to establish. For example, if a young child comes to a caregiver and says that she is sad, a parent does not take the child's claim as de facto truth. Parents often must determine the legitimacy of the tears to decide if they are authentic or if they are, as we say in North America, "crocodile tears" that are put on for show. Billig points out that this social engagement is about training the child to perform sadness in a way that is taken to be authentic. The authenticity of psychological phenomena is something that is established by people and so we do not *have* emotions. Rather, we *perform* them in a way that is culturally satisfactory.

While there are some points of divergence, this view of psychological phenomena resonates reasonably well with one that takes culture to a constitutive quality of mind. Richard Shweder (1991) is a proponent of this view and argues that culture and mind "make each other up." The field of cultural psychology points out that psychological phenomena are shaped through the flow of narratives and metaphors into which we are born. Thoughts and emotions emerge in socio-interactive actions like I described above and so the narratives and metaphors that are integral to this interaction shape possibilities for psychological/performances. This view is highly indebted to anthropological thinking and claims from scholars such a Geertz (1973) that

> Undirected by culture patterns – organized systems of significant symbols – man's [sic.] behaviour would be virtually ungovernable, a mere chaos of pointless acts and exploding emotions, his [sic.] experience virtually shapeless. Culture, the accumulated totality of such patterns, is not just an ornament of human experience but – the principle basis of its specificity – an essential condition for it.
>
> (p. 46)

This is the view on psychological phenomena that influence how I approach research.

What led to me being surprised that I was part of a seminar on the topic of the cognitive science of religion and Christianity was that I take a critical stand toward such disciplines as a cultural psychology. One of the sources of the moniker "cultural psychology" is Shweder, but there are other theorists that support this view. An important figure is Jerome Bruner and his *Acts of Meaning* (1990), where he argued that the cognitive revolution in psychology got it wrong. He made the case that the cognitive revolution in psychology was a reaction against behaviourism and so represented an attempt to focus on meaning in human affairs. The challenge that he raises is that the focus on cognitive processing bypasses the meaning that the cognitive revolution was supposed to recover. Meaning is something constituted in narratives and metaphors that are sociocultural phenomena, and so focussing on self-contained processing mechanisms misses meaning. It is for this reason that he advocated for a cultural psychology in contradiction to the individualistic mode that we see in cognitive science and cognitive psychology. I am heavily influenced by cultural psychology and so found myself amid several researchers that have a completely different view of the research practice of psychology.

I had not applied to study the topic of cognitive science just to spite my colleague because there was a deeper interest. The opportunity to be surprised does not come along as often as good scientists would like, and I think this happens because the pressures of academic labor require one to publish. Staying within one's own domain with colleagues that support one's own work is a good way to generate a consistent stream of publications. What challenged me was my PhD supervisor and my dissertation on Mikhail Bakhtin. I did a project on Bakhtin and how cultural psychology has neglected how the body is an important part of cultural psychology. I argued that culture is deeply embodied and shapes the realities that people experience. Bakhtin illustrates how we see the world in a way where our purview feels irreducibly real. He opens the possibility to talk about dialogue in a context where realities seem irreducible. Dialogue is not just about an interface of perspectives that can be easily relativized and altered. It is about moments when we brush up against the reality of another, and this is what makes dialogue especially tough. It is also what enables the possibility of authentic surprise: dialogue can confront what we take for granted in a deeply existential way. It was this interest in dialogue that really prompted and affirmed my engagement with the cognitive science of religion.

This book is an attempt to enable surprise by escaping the trap of disciplinary thinking. The strong cultural position that I take herein is designed to provoke researchers in the cognitive science of religion. It is an attempt to start from William James to show how the cognitive science of religion bypasses meaning and the specific content of religion. This bypass amounts to a discipline that is good for cognitive scientists but remains largely ineffective in terms of its application in everyday life. I attempt to challenge the common disciplinary thinking that is often seen in the cognitive science of religion. The seminar at Calvin and community of researchers have been very gracious with my critiques and provided me with a body of work that orients me to the discipline.

In the spirit of dialogue, however, I also attempt to push work in cultural psychology in a direction that takes cognitive science seriously. Critical and cultural views of psychology initially started out in a place where they were engaged with mainstream approaches. They have since formed their own communities and have generated their own genres of publication. As an active scholar in the field who has reviewed countless articles and edited some volumes, I have noted a potential for intellectual hubris. I offer a way to think about cognitive science that gives proper due to recent developments in the field.

This book is an effort to be provocative on the one hand. It provokes the cognitive science of religion to think through its presuppositions and challenges weaknesses in the theory. It provokes critically and culturally oriented theorists to take seriously the work that is going on in the cognitive science of religion. Such provocations are what constitute the possibility to build transdisciplinary bridges, on the other hand. Creating a situation of provocation makes enemies on both sides, but it also is what we need to do to engage in dialogue. Confrontation sets up the conditions to see different purviews, and this sort of conversation starter is what this book is intended to be about.

Chapter 1

Toward a cognitive science that doesn't alienate everyone except cognitive scientists[1]

Abstract

This introductory chapter discusses the phenomena that preoccupies this book: religious belief taken for granted as self-evident truisms. A challenge to studying these phenomena is that researchers are "speaking a different language" than the participants in the studies. There seems to be an impasse, and this chapter is about looking to William James as inspiration for how to overcome it. James was against an abstract notion of belief where something can be separated from a belief about the thing. Belief is misunderstood when we separate it from a thing that we have a belief about because belief necessarily involves knowledge about something. He was also against a disembodied notion of belief. Experience includes how our senses entwine with happenings in life. The flow of human experience includes continual relation among ideas and the body, which means that it does not make sense to abstract belief from the concrete materiality of the world. This chapter outlines how an impasse between researchers and participants emerges because the former take an abstract and disembodied approach to belief. The quarrel with research is not with the dismissal of religious belief as an accident, but the underlying presuppositions of what religious belief is. This chapter thereby outlines how the impasse emerged and presents what we can do to move forward.

Introduction

Blake Wenner was a student for whom I supervised a research project on the topic of religious doubt. He was interested in examining how Christian believers developed and reconciled doubts. His desire to do this research came from his own experience where he saw belief held together by dubious rationalizations and thinly veiled self-interest. He recruited and interviewed participants who were Christian believers and what they said, for him, was disappointing.

Instead of participants explaining their rationale for reconciling doubts to fortify beliefs, they often had no clear rationale as to how religious belief was sustained. The participants talked about the dealing with religious doubt as a simple act of belief that just fit with life. For example:

Interview one, page 28[2]

Turn	*Speaker*	
1	Blake	yeah, so um. . . . in light of your questioning of your beliefs and things such as where has Jesus' body gone and, um the ascension of things that aren't quite clear, how are how are these uhm beliefs clarified like how have you resolved. . .
2	Participant 1	I don't worry about em
3	Blake	you don't worry about em
4	Participant 1	no.
5	Blake	so um, you would you would say you wouldn't. . .
6	Participant 1	I trust.
7	Blake	you trust. So rationale, you wouldn't put rationale at the forefront of picking apart those issues
8	Participant 1	oh, I guess I trust that God if he created the world (laughs) then he knows what he's doing

Participants described belief, but they did not spend a lot of time wondering why they believed. They illustrated how belief, like in this quote above, is a simple thing that becomes a matter of trust and not one worthy of preoccupation. It was just taken for granted as a truism, and participants did not know why they believed per se because they simply trusted. Blake did not get what he was looking for, and the question is, "Why?"

Blake is typical of researchers in a way that helps us understand why he did not get what he was looking for. It was like he was speaking a different language than his participants. He was talking about beliefs grounded in careful rationale, and they just didn't talk in these terms. They were talking about a life that simply included belief. While Blake's engagement was kindly agnostic, there are similar instances of researchers or academics whose tone is not so nice. Some academics shout at religious people that they are irrational and ignore evidence (e.g., Dawkins, 2006; Dennet, 2006). Religious people shout back that a researcher just does not "get it." Researchers and believers can come from such different perspectives that they are living in different realities. It is as if the two parties are in different worlds, and my point is that this divergence is the central problem.

This book is about this impasse and how to overcome it by expanding the way we approach the psychology of religion. While there are many kinds of psychological research, I am going to focus in on just one: the cognitive science of religion (CSR; Barrett, 2007). CSR is a field within psychology, but its impact is far ranging as it informs a substantial amount of public and academic discussion (e.g., Ball, 2012; Krakovsky, 2012). CSR is a good approach to engage because it represents the discipline of psychology well as cognition has been identified as central to the discipline of psychology (e.g., Thagard, 2005). As such, I seek to explicate an approach that accommodates both religious believers and psychologists by way of a provocative discussion of CSR. To identify and move beyond this impasse, I am going to look to the past by discussing one of the founders of modern psychology: William James. He had a vision for psychological research that is quite different from CSR, and looking back to him reveals a way of approaching the psychology of religion that can surmount the impasse (see Cresswell et al., 2017). This introductory chapter outlines James' ideas and serves as the context for the remainder of the book. It will first address belief as James described it and then show how it is bypassed in CSR. From there, I will return to James and discuss truth as a springboard for the rest of the book.

William James and the bypass of belief

René Descartes has arguably been one of the most influential thinkers who has shaped our understanding of what things like belief mean. He influenced Immanuel Kant such that they both shared very similar ideas about things like belief (for a sustained discussion see Taylor, 1989). Both Kant and Descartes have left us with a heritage that treats belief as *abstract* and *disembodied* (see Harré, 2002). To refer to belief as abstract is to say that it is not necessarily tied to what is happening in life because it is an ethereal mental property. It involves conceiving of belief as not necessarily tied to the actual happenings of life because it belongs to the realm of subjective mind. To say that belief is disembodied refers to Descartes' famous split between the mind and body. He treated the mind as abstract in its ethereal nature, and this abstractness meant that is was not necessarily tied to the body or anything physical. The body is not tied to belief because it is abstracted from it. William James took a contrary approach.

Belief in relation to something: against abstract belief

James was critical of the notion of an abstract belief (James, 1996/1912). An abstract and disembodied approach to belief involves a separation of something from a belief about the thing. He was against the idea that we can sensibly talk about belief being separate from something in the way that an abstract and disembodied approach implies. Belief is misunderstood when we separate it from a thing that we have a belief about because belief necessarily involves knowledge about something. That is, belief involves knowing about something because we cannot have a belief *about* nothing. The *about* is crucial if we want to understand how belief works in life. Consider the following conversational excerpt as an illustration.

Interview 1, page 15

Turn	*Speaker*	
1	Participant 1	and I I've can't think that you know when you pray, help me to forgive this person
2	Blake	Yeah
3	Participant 1	and love this person, that's so very important
4	Blake	yeah
5	Participant 1	and I have I don't feel that I have a problem with forgiving because I know that there is a god out there that will help me

This is a discussion where a participant talked about a belief that one should pray for someone when that person bothers one. The first use of knowledge shows up in "you know," and this is a common conversation filler (turn 1; ten Have, 2002). Grammatically, it imputes knowledge to Blake and implies something about which there is shared knowledge. When participant 1 says "you know," it is about something, and this thing is the act of praying for someone. The action and potential someone to "pray for" are absolutely necessary for the belief to make sense. Without connection to this wider context, Blake would have no sense of what the participant means. This word "know" is refined in turn 5 to show us how believing in praying to forgive someone involves knowledge *about* God. Turn 5 shows us how belief works in life as connected to a thing "out there" beyond one's own thoughts: God. As we can see in the example above, belief shows up in life in a way that involves knowledge about something "out there" beyond one's own subjective thoughts. James' argument extends to religious belief and shows us how, without something tied to it, religious belief is not sensible. The sophistication of his point lies in how he is not offering apologetics but a way to approach religious belief.

James (1981/1907, 1956/1897, 1996/1912) considered how belief functions in life to show how our religious beliefs are about something. He highlighted how belief is about relationships in the sense that one can have knowledge about another thought or another thing, but it always involves a relation to another thing. All things that fall under the banner of belief involve relationships so that we do not have any element of belief that is isolated and on its own. That is, every belief is related to something else, and it simply does not make sense to talk about a belief that stands as a subjective isomorphic proposition. Belief is a kind of knowledge in relation to something else, and so we always have religious belief in relation to another thing.

This approach to belief is what enables James to break us out of the abstract approach we get from Kant and Descartes, which is important because abstract conceptions of belief simply miss the phenomenon. This position has a huge implication for how we think of ideas like subjective mind and objective reality. James wrote:

> Just so, I maintain, does a given undivided portion of experience, taken in one context of associates, play the part of the knower, of a state of mind, of 'consciousness'; while in a different context the same undivided bit of experience plays the part of the thing known, of an 'objective' content . . . since it can figure in both groups simultaneously we have every right to speak of it as subjective and objective both at once.
>
> (James, 1996/1912, p. 10)

To many people, speaking in terms of belief being both objective and subjective seems strange, but it provides a very helpful idea. A belief in another belief can be subjective, but the web of relations involved in belief never stops there in self-contained subjectivity. All belief eventually comes into relation with something in the world we share with others. Eventually, we come to a relation to something that is not just inside our heads and is necessarily never abstract. Take a close look at the following illustrative conversation.

Interview 2, page 21

Turn	*Speaker*	
1	Blake	. . . do you find that questions are resolved by way of moments of rational clarity
2	Participant 2	are resolved by way of moments of rational clarity did you say
3	Blake	yeah
4	Participant 2	gosh I wish something was totally clear all the time, it'd be nice you know rational clarity to me is easier to see than spiritual clarity and so again sometime although deep down in the bottom of my heart I'm doing something just because it feels right. But again if it does feel right and I have these two equal things I'll go down with what I think is right or what I feel is right I just try to do the right thing
5	Blake	but but
6	Participant 2	and so, and the question about if you give a man a fish he will eat for the day then starve, if you teach him how to fish then he'll live all day but if a guy's hungry I'm gonna give him some fish you know later on I'll teach him how to fish but you know but if he's hungry to me the right thing is to give him some food at that moment. Is it enabling I don't know but it just feels right to me to help somebody

If we look at the structure of this conversation, we see how the participant articulates his belief. Turn 4 involves discussing how something just "feels right," and so it would seem to contradict James by not leading to something outside the participant's head. It reads like a description of a deep solipsistic experience. In turn 6, however, the participant transitions to talking about *doing* the right thing. In the way the participant talks, we see how religious belief ties to an actual activity like helping someone. The structure of the answer reveals a belief in relation to an objective action in the world. It is insufficient to say that the belief is just subjective because it always points beyond an individual to something outside. To paraphrase James (1981/1907), if one dies and stops believing in something, the object does not disappear, and it continues on without any single individual: ". . . I could perfectly well define [belief], what the knowing actually and practically amounts to – leading towards, namely, and terminating in percepts, through a series of transitional experiences *which the world supplies*" (James, 1996/1912, p. 25). Take the example of Participant 1 knowing that there is a God "out there." Obviously, it may seem like God is a subjective thing because we cannot find a conclusive empirical case for its existence. If Participant 1 dies, however, the notion of God is not going to die with him, and so it is not properly subjective. James, contra Descartes, realized that belief does not fit with the subjective-objective dualism that people tend to take for granted.

Belief in relation to something: against disembodied belief

Religious belief, then, is not simply dismissible as a subjective thing, and it is not simply an objective thing. If belief were purely separate from the world and not in relation to something, then we could say that it is abstract. We cannot say so about belief, and so it is not abstract. What about the notion of being disembodied?

Experience involves the body, and so addressing this question involves a discussion of experience (Baerveldt & Voestermans, 2005; Cresswell, 2012; Thompson, 2007). What James (1981/1907, 1996/1912) meant by experience can be understood if we start from basic experience and then work to the more complex notion of religious belief. James goes as far as to claim that belief is best characterized in terms of pure experience, which is the name for a collection of "sensible natures" (1996/1912, p. 27). Experience, at its most basic level, includes how our senses entwine with happenings in life. When one looks out at the environment in which one is reading, for example, light rays stimulate cells on the retina, and the activity of cells happens in relation to what is in life. This is a simplistic example of experience and highlights how experience involves a flow of sensory stimulation occurring in relation to the world. We move through life with our bodies responding in relation to things, and so one experience passes into another in a constant flow.

The constant flow of experience involves more complexity than merely being stimulated by an environment. If one were to sit and read in a café, one could look around and see the stimuli in terms of coherent and meaningful unities like "tables" and "chairs." Previous flows of experience such as learning the names for such things in childhood bleeds into the current flow to give sensory stimulation shape. Included in the previous experiences that are brought into the present are emotions and personal histories. A café may *feel* a certain way because of previous experiences like, for example, spending time with a caregiver at cafés that gives the tables and chairs emotional valence. James (1996/1912) pointed out that experience includes a constant flow of stimuli in relation to one another and in relation to psychological phenomena like concepts and emotions. He pointed out how experience involves an inseparable relation between psychological phenomena and physiological ones. The flow of human experience includes continual relations among a range of elements, and so it does not make sense to abstract conscious phenomena like belief from the concrete materiality of the world. Hence, he argued against the idea that that the "thought-of [an] object is hid away inside the thinking subject" (1996/1909, p. 19) because the past can impact us directly to shape what is experienced as real. Belief is something that is experienced in the flow of the life that we live and experience as real. It is part of the reality that humans experience.

Religious belief is thereby entwined with physical life. Beliefs relate to all other experiences insofar as they are in relation to concrete particulars like light and touch. They involve a complete way of relating to the world that shapes reality that feels objective. James thereby described belief as part of an interconnected web of past and present experiences shaping a reality that includes belief. When someone is faced with a situation, beliefs come to bear in their experiential sense and shape immediate bodily dispositions. Instead of deciding about what to do in life on intellectual (i.e., abstract and disembodied) grounds, we "find ourselves believing, we hardly know how or why" (James, 1956/1897, p. 9).

The question I raise is an important one: Do psychologists, and those involved in CSR in particular, address belief as it shows up in life, in its non-abstract and embodied quality? I don't think so. Blake, for example, was looking for abstract and disembodied rationale for belief that does not include belief as a whole way of experiencing the world. It is an important question to decide if Blake is typical because, if psychologists are giving back to the community like good citizens do, then we ought to have something to say about the actual phenomena of belief as it shows up in life. The next few pages propose why the answer to the question above is partly no.

Psychology of religion: cognitive science of religion and bypassing belief

Dennet's (2006) *Breaking the Spell: Religion as Natural Phenomenon* describes religion as being the accidental result of a "hyperactive agency detection device." "Device" refers to an automatic mechanism that operates in the mind. Humans are supposedly endowed with a mechanism for detecting agency in the world around us. It is "hyperactive" insofar as it kicks into gear and processes all sorts of events as if they were caused by agents. The result is a supposed mechanism that tends to overproduce the perception of agency lying behind natural happenings in the world. The wind rattles a shrub as we walk past and jump back as if there is a nefarious thing intentionally prompting such happenings (see Dennet, 1996).

Other researchers were picked up by the popular media with the claim that "Analytic Thinking Promotes Disbelief" (Gervais & Norenzayan, 2012). Titles like this betray a claim that religion is a product of mechanisms like the hyperactive agency detection device, but controlled rational thinking overriding such intuitive mechanisms leads away from religious belief. People can supposedly override these primal leftovers of the mind when they think more carefully and analytically. The implications of such ideas can be seen in the work of well-known critics of religion like Richard Dawkins: "[t]he general theory of religion as an accidental by-product – a misfiring of something useful – is the one I wish to advocate" (2006, p. 188).

The quarrel with this view is not to quarrel with the dismissal of religious belief as an accident, but the underlying presuppositions of what religious belief is. There is good reason to suspect that this approach, while offering valuable insights in its own right, bypasses religious belief as it plays a role in life. This reason has to do with the way that psychologists have generally approached religion. Usually, they advocate an agnostic approach to religion that considers the topic from a "rigorously" scientific perspective (Argyle, 2000).

At a general level, such a perspective is hard to maintain because studying religion is not an easy task. Hall et al. (2008) pointed out how this challenge is attested to the availability of over 100 instruments attempting to measure religion or aspects thereof. They noted that the challenge is compounded with how the definition as to what even counts as religious shifts with the researchers. In some cases, studies are about religious behaviors such as going to attend a service. Others involve an attempt to measure behaviors and/or more ethereal-experiential aspects of religious experience (see also Kapuscinski & Masters, 2010). These efforts reflect an attempt to get at a sense of connectedness associated with spirituality as opposed to institutional religious behavior or ascent to dogma (see Meezenbroek et al., 2012). What Hall et al. (2008) note is that authors often overstate their conclusions about religion in general when there is so much disagreement as to what the phenomenon is and how it is being measured. Researchers seem to continually cling to the notion that some sort of religion in general must underlie all of this diversity:

> Although the empirical stream of religious measurement has developed and supported a multi-dimensional model of religiousness that resists global assessments, much of the research on religion . . . assumes that "religiousness-in-general" actually exists. As such, it attempts to measure the intensity of religiousness (belief, experience, strength, value, etc.) in order to locate people on a continuum between "very religious" and "not religious" . . .
>
> (Hall et al., 2008, p. 154)

Despite imprecise concepts and a need for contextual social/theological informed research, a context-free approach to religion continues to be popular. It is in this vein that authors such as Underwood (2011) argue that religion and spirituality are multidimensional constructs with many features reflective of a single underlying factor. Developments such as the Daily Spiritual Experience Scale are designed to assess such variability while being opaque enough to even work with people who would not call themselves religions. Measuring experience instead of beliefs is taken to get at this underlying construct (see also Underwood & Teresi, 2002).

The literature points to a recognition that it is difficult to access underlying unified constructs, yet such covering laws must nevertheless be present. Chirkov (2016) notes that this paradigm is one where researchers search for natural laws underlying behavior. That is, researchers are interested in uncovering the truth about reality in a way that has been referred to as naturalism (see Reber, 2006; Brown & Stenner, 2009). Naturalism involves trying to understand the truth of reality by uncovering the laws by which the natural world works. This approach is that one has found truth when one finds the mechanical laws by which the world works.

In the early 1960s, psychologists began to speculate on how the mind worked in order to uncover the naturalist laws of mental operations. Alan Turing – one of the widely accepted forerunners of the cognitive revolution that shapes contemporary psychology (see Dawson, 2001) – developed the idea that humans operated on computational principles, where computation was defined as algorithms or rules (Boden, 2006, p. 173). This person-as-computer perspective put forward by Turing was that the computer observed symbols, processed such symbols according to a set of rules, and then responded on the basis of the outcome of those rules. One author writes that "[c]ognitive skills are realized by production rules" where "[p]roduction rules are if-then or condition-action plans. The if, or condition, part specifies the circumstance under which the rule will apply. The then, or action, part of the rule specifies what to do in that circumstance" (Anderson, 1998, pp. 59, 63).

It is on this heritage that much of contemporary psychology is built (see Thagard, 2005). When authors argue for an agnostic scientific approach, they mean a science that uncovers natural laws about mental computation and outputs. It involves seeking to explicate psychology in terms of natural mechanical laws of human functioning. The result is an approach to religious belief shaped to fit a naturalistic bias. The rigorous scientific approach is a form of naturalism that shows up in in the psychology of religion as a concern with computational processing laws.[3] Discovering naturalist predictors and processing outcomes associated with religiosity reflects a concern with what psychology is taken to be about. This kind of approach is what is embodied in a "rigorously" scientific perspective. Nowhere is this heritage clearer than in CSR. The sort of production-rule-based naturalism manifests in the CSR, and writing about practices in this field can be generalized back to the psychology of religion in general. CSR, for example, underlies the work on the Hyper-Active Agency Detection Device that I discussed above (e.g., Guthrie, 1993; Whitehouse, 2004).

This kind of naturalism bypasses the phenomena of belief as it was outlined above in our discussion of James. An approach that is scientific via a naturalist fixation can come at the expense of everyday life (James, 1981/1907). Consider as an example how some authors of the naturalist ilk have explicitly stated that they are not interested in what belief is about because this is simply not a relevant focus (e.g., Pyysiäinen, 2002; Guthrie, 2002). That is, belief as a non-abstract and embodied phenomenon is not a concern. Religious belief, as it shows up in life, is part of the reality humans experience and is not lived as a cognitive mechanism. Consider an illustration from Blake's interviews.

Interview 5, page 20

Turn	*Speaker*	
1	Blake	yeah, uhm. so why would you say just expanding on that, that you rely more on your feelings or intuition
2	Participant 5	uhm, . . . well, it's not, for me it's not fleeting, it's not uhm, wispy or insubstantially. That stuff comes off really powerfully
3	Blake	Mhmm
4	Participant 5	with me and part of that's being so harmed when I was a child, those intuitive things go way up
5	Blake	Mhmm
6	Participant 5	it's like God somehow uses your brain, to pay attention in ways when you've been harmed, that doesn't happen with other people and probably has to do with you know brain development
7	Blake	yeah
8	Participant 5	it's the stuff that's I think of it now. So it's not wispy it's not insubstantial it's really powerful

This quote would seem to support the idea that a cognitive mechanism is at stake with the mention of brain development. This participant, however, is talking about the experiential power of belief. God is described as part of reality by the implicit presupposition that God uses one's brain. A proponent of CSR would say that this experience is irrelevant because it only matters what happens underneath in the realm of naturalist mechanisms. This position is fine for cognitive scientists who believe – in the full sense of the term – in naturalist laws.

Belief, on a naturalist CSR approach, is abstract in the sense that the specifics that belief is about are simply not addressed. The hyper-active agency detection device, for example, allows for any kind of super-agent to fall into the scope of the mechanism's operation but a belief is rarely about a non-specific super-agent. For example:

Interview 2, page 8

Turn	*Speaker*	
1	Participant 2	so now and at this one point yes I believe in a virgin no I don't no I didn't yes I do no I don't. I think that my faith is now at the point that if there was something that came along that proved scientifically that Jesus was conceived by Joseph that it wouldn't affect my faith at all because again all things are possible through Jesus . . . it just isn't that matter now that would be a serious problem for a lot of Christians to believe that Jesus wasn't from a virgin relationship but that wouldn't be a problem form me

Interview 2, page 8		
Turn	*Speaker*	
2	Blake	Right
3	Participant 2	disproving the resurrection would be cause I think that's pretty essential

Interview 3, page 3		
1	Participant 3	there's an enormous amount of opportunity to think about the experience of Christ and his life with an accent on ah, his suffering and his role as the redeemer

In both of these examples, the beliefs are about specific features of the faith. What practically matters in life is what beliefs are specifically about. A general mechanistic rule is too abstract to grasp what this belief entails. The approach taken by CSR is also disembodied insofar as experience is bypassed. Moreover, a specific religious belief involves particular content that is integrated with emotional entanglement with the world. The abstract mechanisms bypass this integration and lose what it means to have the experience of religious belief.

For those outside of the CSR belief system, naturalistic laws of functioning are unhelpful because the psychological reality that matters in life is bypassed. When truth is defined as only finding natural mechanisms and processing rules, we actually bypass what religious belief is. The result is that belief, as it tends to be studied, substantially diverges from the way it shows up in life. It is here that we can see the impasse that Blake and other researchers reach. The dismissal of belief as a processing mechanism is met with a response that what is dismissed is what really matters.

Moving forward

James shows us how religious belief works in a way that is concrete and embodied. A related issue is that how we think of the scientific belief can be abstract and disembodied. Naturalism in CSR exemplifies this issue because a claim in CSR is taken to be true when it represents what natural mechanisms are at play. For example, if a researcher claims that the hyper-active agency detection device produces the ability to perceive a super-agent, a researcher is making the claim about a mechanism that plays a role in the production of religion (e.g., Boyer, 2001). It is taken to be a true claim because it represents a natural process that is really happening. CSR, for example, is abstract in the way that naturalism bypasses what belief is about. A super-agent and experience that a religious belief entails is treated as irrelevant when scientific truth is approached in this manner. It is disembodied in the sense the experience of life is little more than a source of stimulation. This bypass of belief comes from an approach to science where truth is abstract and disembodied. The truth claims in CSR are about general naturalistic mechanisms that don't bear much relevance to religious beliefs as they show up in life.

James (1981/1907, 2011/1909) was aware that holding only to a naturalist approach to belief means that it is often bypassed and addressing religious belief requires a broader approach to scientific truth. That is, instead of presuming natural cognitive mechanisms to be the only feature of belief that matters, we can return to James because he offers a direction that expands the notion of scientific truth. This section outlines a vision for CSR that sets the groundwork for the future steps proposed herein. For our purposes, I will outline three criteria for truth that enable this expansion.

First criterion for truth: fit in life

James (2011/1909) expanded the definition of truth by arguing that it is always inextricably bound to life because people do not use truth in an abstract and disembodied sense. This means that truth is not used in a vacuum because people use it in relation to all of the other aspects of life. For example, consider the following interview extract.

Interview 3, page 1

Turn	*Speaker*	
1	Participant 3	where even when I was very young I was participating in a, one could say a special way and so my life was uhm . . . steeped in religion literally from morning to night, (laugh)
2	Blake	mmm I see
3	Participant 3	and then uhm, when I became an adult I changed dramatically I came back to the religious community fairly recently. so I would say most of my adult life I was not a practicing Christian

To understand the notion of truth in order to study religious belief, we need to look carefully at this conversational excerpt. This participant discusses leaving faith and coming back to it. Religious belief is discussed in turn 3 in terms of being a "practicing Christian": doing something in life. We see the theme of religious belief being tied to embodied doings and not a subjectivist ascent. It also refers to life being steeped in religion and coming back to a place where Christianity is considered true is coming back to *practice*. Here we see a discussion of the idea that religious belief is tied to practice, and what I want to add is that truth is tied to life outside of the mind. Consider this idea in light of the next illustrative comments.

This participant talks about belief in relation to aspects of life such as all of the vicissitudes of life growing up. Religious belief emerges in the interconnected milieu of life and so it must be part of life to be true. A religious belief is true in the context of interrelations among emotional valuations, ideas, things, and life with others. These interrelations all play a part in saying that a belief is true. Truth is not just separate from life in the abstract and disembodied sense, and a belief is true because of how it relates to emotional valuations, ideas, others, and things that people are engaged with. Namely, it has to fit together with the rest of life.

It is the interrelations among features of life that provide a criterion for how we can approach scientific truth. This criterion is that something is true when it resonates with the rest of life. We live in an interconnected experiential web of things and ideas (past, present, and anticipated) and this interconnectedness extends to a scientist. James' view of truth was that a scientific claim is true when it fits into the flow of life insofar as it fits into the complex web of interrelations constituting experience. There is a sense of satisfactory peace and rest when something fits with lived experience. For example, reconsider a quote that I presented earlier. Consider what we see in this participant and what it can tell us about the application of science in CSR.

Interview 2, page 21

Turn	*Speaker*	
1	Blake	in light of your questioning your beliefs do you find that questions are resolved by way of moments of rational clarity
2	Participant 2	are resolved by way of moments of rational clarity did you say
3	Blake	yeah
4	Participant 2	gosh I wish something was totally clear all the time, it'd be nice you know rational clarity to me is easier to see than spiritual clarity and so again sometime although deep down in the bottom of my heart I'm doing something just because it feels right. But again if it does feel right and I have these two equal things I'll go down with what I think is right or what I feel is right I just try to do the right thing

Consider also:

Interview 5, pages 1–2

Turn	*Speaker*	
1	Participant 4	. . . my life was tough in that my mother died when I was seventeen, and there were money problems although my dad didn't talk about it much but, there wasn't much money. and uhm, managed to get an education at [university name] and uhm . . . my religious beliefs and practices . . . have arisen quite naturally out of the way a life develops

This quote illustrates how the truth of religious belief is substantiated when it "feels right" "deep down," and it illustrates a sense of peace and rest being a criterion for truth. A religious belief that seems disjointed from the aspects of life would not be true and satisfying because it simply does not fit with life. The same is true for a scientist who finds truth when it fits with life. It is the different realities lived by a religious believer and a scientist that provide different contexts enabling different claims to be true. A naturalist CSR approach may fit for those that believe in it, but it runs the risk of stopping there. A religious belief is taken as true when it fits with the experiences that are part of a different form of life, and we can see how a naturalist concern with mechanisms would not fit. At the moment that it fits the flow of experience, whether it be that of a believer or that of a scientist, belief is experienced as true. A claim is true for CSR researchers when it fits with a reality that includes naturalism. A claim made by a scientist in CSR gains the status of truth beyond the reality of scientists the moment that it falls into resonance with the reality of those to whom she is proclaiming a truth.

In the next chapter, I elaborate the forgoing to outline an approach to research that can get at religious belief in a way that fits with the world of religious believers. A problem raised above is that an abstract and disembodied approach to belief misses what belief is and the constitution of different realities entwined with it. One key to getting at what religious belief is about is to take seriously the role of culture because cultural narratives are often central to the constitution of such realities. CSR has yet to develop an account of culturally shaped religious belief because of its naturalist approach that is unconcerned with culturally specific beliefs (Cresswell, 2014). Culturally shaped beliefs can potentially – and inappropriately – be reduced to stored information, and experience is only understood as an accidental by-product of computation. Chapter 2 thereby addresses how culture is necessary for a discussion of belief if one is to find a fit with culturally shaped life: to truly understand belief is to grasp the culturally shaped stories that are tied to it.

Second criterion for truth: fit with a community

To get away from the problems associated with abstract and disembodied truth, it looks like objectivity is lost. This implication is correct, but a collapse into subjective fancy does not necessarily follow. James (1981, 2011) repeatedly argued that truth is not a matter of the mere proclivities of subjectivity. He did so in a way that points out how one's truth is only true when it resonates with life outside one's mind and so means that it resonates with others. He made the claim that standards of fit are socio-normative in quality and are integrally integrated with life with others. People's precepts are common and they are such within a community. The truth of something ". . . gives us an absolute phase of the universe. It is the personal experience of the most qualified in *our circle* of knowledge to have experience, to tell us what is" (James, 1981, p. 17). For James, the truth of belief is deeply entwined with living with others with whom we agree on what counts as true. For example, even when someone makes the claim that one must find one's own truth as in the following:

Interview 6, page 1		
Turn	*Speaker*	
1	Participant 6	. . . ah not so much any, you know really earth shattering event, just . . . just like as I grew up uhm, I just noticed ah the people I associated with and their lifestyle and how they were affecting me more than anything
2	Blake	mm
3	Participant 6	and there's no, no big light event or anything like that. . .

Participant 6 is talking about the impact of others and highlights how important others are for belief. There is not often a decontextualized moment of divine revelation that cannot be brought into context of relationships with others and the world. One cannot have a true belief on one's own. Beliefs have truth by virtue of fit and generativity within communities. While not objective, the truth of a belief is certainly not subjective because it cannot be reduced to an individual unit.

In short, one cannot have a belief that is true on one's own because it must fit with communal standards. This idea can be extended to scientific truth as well. It is not necessarily objective, but it is certainly not subjective. A true belief is one that fits with the standards of a community, and because there are different communities in the world, there are different standards of truth. There is a long tradition of work showing how the standards of naturalist scientific truth are normatively grounded (see Danziger, 1990; Gergen, 1985; Stam, 2015a; Brown & Stenner, 2009). Consider a simplistic example from the history of psychology. During the Behaviorist era, standards of truth revolved around the study of behavior without reference to the mind. Standards of scientific truth then changed in the 1960s so that research excluding the happenings of internal mechanisms is now considered dubious. The truth about naturalist mechanisms in CSR now works for that community, and it is time for a broader approach that speaks to a community beyond its own.

Where Chapter 2 specifies how we can reconsider religious belief from within the purview of religious believers, Chapter 3 offers something palpable to researchers in CSR. It presents how child development involves learning to embody communal standards. Babies usually experience emotions as intense global emotional states that result in undifferentiated crying. Such dispositions are cultivated into normative signals as caregivers socialize children by their responses to children's emotions, enabling the transition from global catastrophic emotion (e.g., crying) to emotional signals (e.g., crying to obtain). Such interpersonal experiences become the bedrock of cognition where early childhood interactions are also sites of the development of socioculturally constituted cognition. This chapter thereby addresses how we can study the cultivation of cognition in infancy so as to show the development essential for religious belief and the meaningful experience it entails. Since such cultivation of cognition involves emotional signaling and the importance of interrelationships, it speaks to my goal of developing an approach to cognition that includes socially shaped religious belief. On the one hand, Chapter 3 continues with a discussion of the communal standards that constitute the truth of a religious belief. On the other hand, it outlines the kinds of changes necessary for researchers to broaden our science by providing an approach to cognition.

Third criterion for truth: generativity

Another criterion that establishes truth builds on previous ones insofar as fit with life is not just a static fit with other aspects of experience at one moment of time. Humans are always growing and changing in dynamic ways. We change and the world around us changes such that a true belief could be antiquated and irrelevant. A true belief, James (1981/1907, 2011/1909) argues, is one that fits the dynamics of life to do more. For a belief to remain true, it must generate something for us as we move forward such that it makes future activities smoother. Truth is "essentially bound up with the way in which one moment of our experience may lead us towards other moments which it will be worthwhile to have been lead to" (1981/1909, pp. 93–94). A true belief helps us succeed in life as opposed to creating friction and tension. Consider the following conversational illustration.

Interview 2, page 29

Turn	*Speaker*	
1	Participant 2	I don't know (laugh)
2	Blake	hmm, so uhm if you guys were having a conversation it wouldn't really get anywhere like, because his beliefs are kinda there, and yours are kinda here
3	Participant 2	yeah I get the impression that we would have a very . . . I think I'd have a satisfying conversation I don't think he would bring me along or I would bring him along but I think we'd leave friends
4	Blake	yeah so its still
5	Participant 2	so
6	Blake	but you'd still be able to reconcile differences
7	Participant 2	yeah yeah, well there would still be differences and we'd just live with each other

This segment involves a description of a discussion Participant 2 had with someone that has a radically different perspective. Participant 2 and the conversational partner he describes do not resonate with each other in terms of the religious beliefs that each hold true. They cannot reconcile and cannot move forward into a new and deeper relationship, so it halts. When ideas or claims interrupt the flow of life, we then examine their truthfulness or we move to another group or context where beliefs allow us to move forward. A discontinuity between the two beliefs do not mutually work as true, and so the only option is to be pragmatic and move forward in different directions. Their beliefs would work with others that share the same suppositions, and so Participant 2 and this acquaintance must find others with whom their beliefs work if they want to move forward. Participant 2 doesn't examine the truthfulness of his own beliefs and so moves forward to deepen different relationships.

The same criterion can be applied to scientific truth insofar as it remains true when a claim helps people move forward in a more seamless manner. CSR researchers have been moving forward together insofar as their work has generated lots of ideas and discoveries. Unfortunately, there are times where religious believers and researchers have come to a place like Participant 2 and his friend: things aren't going anywhere, so we should "just live with it" by going down our own paths. The problem is that it has not been generative for religious believers due to abstract and disembodied naturalism. Chapter 2 shows how belief can be studied in a way that is generative for believers. If we stopped our discussion at the end of Chapter 2, we would be at an impasse that does not resonate with cognitive scientists. Consequently, Chapter 3 is about how CSR can be a more robust enterprise. The goal herein is to broaden the way we approach the study of religious belief so we can move forward in a broader range of circles. An approach to truth that can fit for religious believers by looking at belief as concrete and embodied moves us along.

A challenge lies in creating a point of generative interface between the strong cultural approach outlined in Chapter 2 and develop an approach to cognition that I initiate in Chapter 3. Chapter 4 further explores a unique approach to the evolution of human cognition with the aim of providing an approach that better includes the cultural quality of religious belief outlined in Chapter 2 and articulated in Chapter 3. Current work in CSR and evolutionary psychology is primarily concerned with genetic inheritance of mechanisms. Evan Thompson (2007) can aid this exploration because he presents a phenomenologically oriented view of the evolution of human cognition that is *fundamentally interdependent* with the surrounding socio-cultural milieu. This is a unique approach that treats religious belief in terms of the whole sensory-motor activity of the body as it extends beyond the brain into the whole environment. Rather than treating genes as the evolutionary unit of transmission, this vision of cognition sees the unit of evolution as the selection for and passing on of dynamic cultural belief systems, thereby enabling religious beliefs to be studied as relational phenomena. This work thereby expands our understanding of human evolution by highlighting how religious belief emerges in interdependent dynamic cultural systems. This is an approach to cognition that is not abstract and disembodied while being generative via its theory of cognition that can account for religious belief. As such, it helps to build a generative bridge between CSR and religious believers. Coupling this work with CSR enables a conception of the evolution of mind that accounts for the universal character of some cognitive capacities underlying religious experience without falling into a view that bypasses belief.

Wrapping up

The concluding Chapters 5 and 6 attempt to consolidate the forgoing. Chapter 5 outlines a vision for how psychologists ought to go about their research of cognition by discussing the notion of virtue. That is, a CSR that does not alienate itself from religious believers should include standards of good research. I will outline such through a discussion of virtues needed to be a good researcher. In terms of the concluding Chapter 6, the approach above does not legitimate authoritative belief taken as a transcendent universal truth because it cannot be determinately grounded in a transcendent being. I propose that this is not a problem insofar as a culturally informed CSR that speaks to both social scientists and religious practitioners is possible. I illustrate how the indeterminacy introduced by the forgoing is not threatening to religious experience. The kind of certainty that comes from appeal to metaphysics paradoxically deadens a believer's experience of God because it leads to faith and commitment to the authoritative rules instead of a living and dynamic engagement with God. The deconstruction of authoritative experience that is part of the proposed approach makes room for dynamic personal experiences.

Notes

1 This chapter emerged in discussions with Blake Wenner and some of the ideas herein were presented at a conference for the Society for the Scientific Study of Religion (Cresswell, et al., 2014)

2 There were six interviews overall, and they were conducted with permission of a local university ethics review board. To protect participants' anonymity, they will be simply referred to as "Participant #." The text was transcribed to be close to everyday speech instead of "cleaning up" the grammar to make it like written text.

3 Of course, I am making general claims, and there are very sophisticated models of cognition that address cognition as non-representational and non-computational (see Neisser, 1994; Shapiro, 2011). The problem is that such sophistication does not generally emerge in the psychology of religion, and this topic is more carefully addressed in Chapters 3 and 4.

Chapter 2

Seriously, culture and language need to be taken seriously

Abstract

James noted that truth is always inextricably bound to life because people do not use truth in an abstract and disembodied sense. Naturalistic explanations of the sort we see in the cognitive science of religions risk bypassing the meaning and experience entailed in religious belief when they bypass what belief is about in favor of cognitive mechanisms. This chapter is about developing an approach to scientific truth that avoids this bypass, which means developing claims that fit with the lives of believers. A culturally informed approach is necessary to accomplish this task. I provide a brief exploration of the notion of culture in psychology and anthropology, since both these perspectives come to bear on the cognitive science of religion. I argue that culture involves meaning systems and cannot be reduced or equated with the notion of nation or broad categorical variables. Detailed understandings of meanings wrapped up in language and narratives, are important, and researchers ought to avoid causal stories without these meanings. I thereby take an approach to culture that is antagonistic to the idea that it is layered over top of more fundamental aspects of humans. Such a culturally informed approach is ideal because it does not treat the meaning of religious belief as secondary.

Introduction

To move forward in a CSR that accounts for religious belief, the criterion of fit with life must be addressed. This criterion is a challenge because it involves understanding religious belief from the perspective of those who live it, as opposed to researchers' a priori presuppositions about the subject matter. William James drew attention to the importance of understanding belief from the perspective of the people who live it. In his seminal *The Varieties of Religious Experience* (1982/1902), he pointed out how religious beliefs are often relegated to a "nothing but –" and the previous chapter addressed how the kind of naturalism we see in CSR could result in treating belief as nothing but cognitive mechanisms. James' position was that naturalistic explanations of that sort bypass the meaning and experience entailed in religious belief when they bypass what belief is about. Consider the following excerpt:

Interview 5

Turn	*Speaker*	
1	Participant 5	then my cousin, who is the woman up by the desk, got a very serious cancer diagnosis; stage four plus plus plus and I'm thinking, we didn't know if she was gonna live . . . and ah, we all sorta went into shock, and I decided that the one thing I could genuinely do with an open heart it was some commitment of time and practice, was to pray for her in a very solid way, and that was about an hour a day, flat on my bed begging for her life. Begging for the medicines to work, begging for the community support we needed, and all of that came through in the gentlest most sweet most supportive way possible and the thing I learned from that experience was that God doesn't choose to heal or not, God is healing
2	Blake	Mm
3	Participant 5	in his presence you heal
4	Blake	Yeah
5	Participant 5	so the trick was to get into his presence
6	Blake	mmm, and prayer was
7	Participant 5	and prayer was the tool for that.

A CSR approach to this prayer could be irrelevant to the lives of believers, and it would be less true relative to a theory that enables a believer to apprehend more fully why she prays and what it means to her. The challenge is to find a way to stay faithful to the life lived by believers instead of turning to a reductionist explanation. This orientation introduces truth criteria requiring a researcher to defend his claims' relevance in light of the meanings and experiences of those he seeks to understand. James' point was that the endeavors of psychologists have little utility in the lives of those we seek to understand unless everyday meanings and experiences are accounted for (James, 1981/1907, pp. 123–134). It is possible to study religious belief and simultaneously have nothing to say about how it works in everyday life.

In contrast, "[w]hat you want is a philosophy that will . . . make some positive connection with this actual world of finite lives." (James, 1981/1907, p. 9). James (2011/1909) expanded the definition of truth by arguing that it is always inextricably bound to life because people do not use truth in an abstract and disembodied sense. Truth is not used in a vacuum because people use it in relation to all of the other aspects of life. Truth pertains to the fit in life and so a true theory, when conceived as James argues, fits common sense and makes common meaning and experiences more fully understood. We are looking for an approach to religious belief that fits with life insofar as scientific truths make some practical difference (e.g., James, 1981/1907, p. 18). Such an approach is quite radical in the sense that common sense is often considered problematic (e.g., Bering, 2006), but later chapters will address how this need not be the case – we do not necessarily fall into fanciful speculation without accountability. A concrete and embodied approach to belief means that we need to address the specifics of what belief is about (i.e., meaning), and we need to address experiences that are part of belief. This chapter is about developing an approach to scientific truth that leads to claims that fit with the lives of believers.

To do so, I will present an approach entitled *cultural psychology* that is especially focused on meaning by way of how it attends on the content of language and stories that people tell (Bruner, 1990; Shweder, 1991; Valsiner, 2012). That is, it is an approach that is specifically interested in what people talk *about* and so fits well with an approach to belief. As I will present below, it is also an approach that is very helpful for understanding belief as something tied to practice and how the apprehension of truth is also tied to life outside of mind. Cultural psychology is particularly adept at articulating how a belief is true in the context of interrelations among emotional valuations, ideas, things, and life with others. James' view of truth was that a scientific claim is true when it fits into the flow of life insofar as it fits into the complex web of interrelations constituting experience, and cultural psychology is interested in this complex web. It resonates with the idea that truth is not just separate from life in the abstract and disembodied sense. As such, it attends to the interrelations among features of life that provide a criterion for how we can approach scientific truth: resonance with life. I first need to set the context by discussing the topics of culture and psychology.

A brief interlude on the topic of culture and psychology

Cultural psychology brings together culture and psychology. Most of this book involves exploring the cognitive science of religion, but it is important to contextualize the notion of culture and what I mean by it. Consequently, I need to review a little about the concept of culture and how we can incorporate it into a discussion with pragmatic value to religious believers. My discussion will be a brief exploration of the notion of culture in psychology and anthropology, since both these perspectives come to bear on the cognitive science of religion.

In the 1970s, cross-cultural psychology emerged out of work done by psychologists who also had an interest in anthropology (see Segall et al., 1998). Scholars like Berry, Bond, Smith, and others noted that there was a lack of concern for culture in the psychological sciences. These researchers were pioneers in pointing out to psychologists that the influences of culture on psychological phenomena are important considerations. The name of the discipline emerged from an attempt to show that psychological phenomena could vary cross-culturally. Initial studies took a cross-cultural approach by focusing on testing people in different locations. For example, Hofstede's (1980) famous studies on cross-cultural values was executed by comparing people who all worked for a single company located in several different nations. Consequently, some of the earliest work in the fledgling field was about comparing people from supposedly different cultures on some sort of psychological measure. Not much time was devoted to understanding what was meant by culture.

It was some time later that the notion of culture began to be problematized (Rohner, 1984). The notion of nation was often equated with the notion of culture for convenience's sake, but this equivalence was recognized as being problematic. A nation cannot be easily defined as the boundary of a culture and vice versa. Rohner's (1984) article was an attempt to bring conceptual clarity to the notion of culture and so enhance the heuristic value of the concept in cross-cultural psychology. Culture, he argued, does not have material substance because it is an ideational set of meanings that could be found in a people group and transmitted over generations. Rohner's position was sophisticated because he noted that ideational systems demarcating culture could not be described merely in terms of sharing or uniformity. The generic patterns can be understood, but not every member of a culture would be uniform in an essential set of properties, and people in a culture do not have a set of a priori shared postulates. Overall, he treated culture as an ideational pattern among a group of people with something like a family resemblance: all generically similar but with no single essence uniting them all.

Historical accounts of cross-cultural psychology show that this articulation did strikingly little in terms of changing practices (e.g., Chirkov, 2009). Culture continued to be used as a categorical variable where nation stands in for culture and ideational meaning systems, as such, were rarely discussed. A decade and a half after Rohner (1984), researchers returned to the effort of articulating what was meant by the concept of culture. Smith and Bond (1999) put out a social psychology textbook with an attempt to offer a cross-cultural approach and argued that culture remained a neglected concept in psychological research. An innovative move by Smith and Bond was to note that culture can sometimes be treated as a pervasive theoretical cure-all. Calling culture a system of meaning and symbols does not help when researchers do not know how to measure or assess such systems. It can serve as a category that can be invoked to explain behavior with very little understanding about what culture is that it can cause behavior. Of course, it can be an explanation for behavior, but they noted that culture can be so broad that psychological theory about causal linkages can become mystified: using culture as an explanation for behavior without a deep understanding can lead to a false theory of causation.

Kashima (2000) tried to demystify the research and understanding of culture by revisiting previous definitions a year later. His article put together an emerging consensus that revisited the notion of materiality by grounding his discussion in Darwinian evolution (see Chapter 4). The important claim was that human evolution involved culture and so our biology is not entirely separate from culture. The broad claim was that the cultural-historical context and the human mind are mutually constitutive. Kashima argued that culture as a meaning system relates to the long-term and continued transmission of similar meanings. In the comparatively immediate short term, however, the embodiment of such meanings is seen in the signification practices. Formulating culture in this way allowed Kashima to point out that information is transmitted genetically across generations and transmitted symbolically within and across generations. Kashima therein tied culture as ideational systems to embodied interactions and reaffirmed this position when he returned to the topic sixteen years later (Kashima, 2016). He added the claim that persons are physically realized, evolutionarily endowed, and culture-historically complemented to make a case that culture is part of human phylogenetic history. In his effort to naturalize culture, Kashima placed further emphasis on what he considers ecological correlates of culture such as body, brain, genes, and ecosystems. Overall, he argued for the importance of the body and warned against treating culture only as a matter of extra-somatic adaptation to the environment.

Cross-cultural psychology pointed out that culture involves meaning systems but that it is necessary to grasp the located meanings at play in cultures. It is important to not just treat it as broad categorical variables. Detailed understandings of meanings are important, and researchers ought to avoid causal stories without these meanings. This view also shows how culture is not entirely separate from the body or materiality. It's worth noting that CSR picks up on these same themes of materiality and naturalism.

The approach to culture taken by CSR, in large, is best represented by an "epidemiological view" (see Sperber, 1996). This view is an attempt to articulate what culture is by not taking a reductionist view where researchers do not seek to reduce it to solely biochemical interactions. They also do not want to divorce culture from materiality in some sort of superorganic conceptualization. Many such researchers look to Sperber's (1996) argument that materiality is important insofar as culture entails bodily movements and physical interaction. This position means that culture is not autonomous from humans and human actions. The epidemiological view purports that there are different interdependent ontological levels at stake and the materiality of culture is one of them. The most pertinent ontological level has been described in terms of memes: cultural elements that are selected to be passed from one generation to the next (see Richerson & Newson, 2009). Memes can be passed from one generation to another in a manner that is parallel to genetic transmission, but on a symbolic level. The kinds of memes that are of interest to CSR are what researchers call "culturally-postulated super agents" (e.g., McCauley & Lawson, 2002). These are agents that supersede human capacities by possessing traits that normal humans do not have and they are concepts (i.e., memes) passed from one generation to another. The idea is that children are born into cultures with concepts of such super agents and to religion is treated as a cultural system that is distributed through a population (see Whitehouse, 2004).

To understand how this orientation to culture can be called an epidemiological view, one must first understand how researchers in CSR naturalize culture through their understanding of how it relates to cognition. Fundamental to the approach to culture in CSR are scripts or prototypes conceived in terms of representations (e.g., Bering, 2010; Boyer, 2001; McCauley & Lawson, 2002). These are concepts that are often shared, and common ones would include a representation of what agents typically do and do not do. A representation of a regular agent may be that it cannot walk through walls because it has a corporeal body while a super-agent may be able conceptualized as being able to do this. Other kinds of representations can be heuristics that we take for granted, such as a common-sense dualism that people have a corporeal body and an ethereal spirit. An important point is that such representations are tied to intra-individual processes and inter-individual processes and so Sperber (1996) refers to public and private representations (see also McCauley & Lawson, 2002). Public representations are available in the public realm to be identified and described by researchers. They can be catalogued and interpreted because they are not necessarily individual insofar as a community will often hold them in common. These are the cultural representations that are predicated on private representations. Humans' individual subjectivities are full of representations of public representations, which amounts to minds filled with representations of cultural objects (see also McCauley & Lawson, 2002).

Researchers in CSR argue that some representations are more likely to be retained than others. Humans are endowed with a "psychological susceptibility" (Sperber, 1996, p. 57) to cultural representations insofar as we have cognitive mechanisms that enable concepts to be retained. Concepts that fit what we are cognitively prepared to acquire are the ones that remain and become cultural representations. For example, Barrett (2008) discusses how we have cognitive mechanisms that aid in the retention of culturally postulated super agents. Such agents that have a few counter-intuitive properties but are still like humans and tend to be retained as enduring representations relative to agents that are either mundane or have many counterintuitive properties. Humans have genetically prespecified cognitive architecture that allows religious concepts to be retained and passed from one person to another (see Whitehouse, 2004). That is, "the human mind has a suite of cognitive mechanisms that collaborate in specifiable and predictable ways to generate religion pan-culturally" (Murray & Goldberg, 2009, p. 184).

It is because of the foregoing that researchers in CSR put forward an epidemiological view of culture. They argue that culture is made up of contagious ideas and productions that are contagious precisely because they fit human architecture (see Sperber, 1996). The combined effect of many small mechanisms enables a wider pattern of cultural representations to be retained. That is, the memes that run throughout social life do so because human cognitive architecture is predisposed to catch them from others. The transmission of notions like culturally postulated super-agents happens on the basis of a cognitive readiness to pick up some ideas and not others.

The notion of memes fits with the meaning system addressed in cross-cultural psychology insofar as a meaning system is a system of memes. CSR focuses on the cognitive architecture that enables specific meanings to be transmitted and retained, which means that CSR focuses on specific components of a meaning system. From CSR, we get an emphasis on cognition and the enablement thereof. From cross-cultural psychology, we get the importance of grasping such meanings. The embodiment of culture is presumed in the relationship between brain architecture and cognitive mechanisms.

Both genres of scholarship are situated in an intellectual heritage from cultural anthropology. Keesing's (1974) landmark essay in theories of culture surveys anthropological approaches to culture that discuss themes identified above. He reviews the way cultures can be understood as adaptive systems where culture is an adaptation to ecological niches, which emerge via natural selection. This discussion opens the notion that culture involves ideational components with adaptive consequences and is resonant with the foregoing discussions of meaning systems and memes. It highlights how culture can be approached in terms of "ideational theories" that treat culture as a function of cognitive systems and so connects with CSR. Such systems include ideational codes in the form of symbols that entail shared meanings. The shared quality of such symbols, in turn, enables researchers to address the issue of sociocultural systems. Patterns of life, argues Keesing, amount to represent ideological enactments of individual designs for living rooted in such cognitive systems. He emphasized how culture therein operates as unconscious systems of knowledge stored inside people's minds and picked up on such ideas twenty years later (Keesing, 1990). Therein, he noted that it is important to recognize that consideration of ecological settings points to the need to think about culture in terms of small-scale communities (Keesing, 1990). Researchers can overstate differences between nations and become tied up on large scale speculations about differences in a search for new and exotic cultures. Missing how addressing culture in terms of distinguishing "theirs" from "ours" is tied into description of symbolic systems of class and power can cause researchers to lose sight of the importance of communal- and interpersonal-level embodied interaction.

While Keesing's (1974, 1990) work aligns reasonably well with CSR and cross-cultural psychology, Geertz (1973) would seem to be an equally important figure that raises opportunity for new directions. Geertz's (1973) famous anthology of essays entitled *The Interpretation of Cultures* involves an attack on what he calls the "stratiographic view" of persons. This view is that humans are made of many levels starting from the most basic level of bio-chemical interaction and extending to the most abstract level of super organic (nearly ethereal) culture. The strata that supposedly make up humans can be understood as somewhat independent domains that interact to produce the phenomena of humans. He identifies how this view tends to prioritize basic levels and treats them as fundamental building blocks upon which successive strata are built. In contrast, Geertz argues that culture is not added onto humans but is part of production of humans. He writes:

> Undirected by culture patterns – organized systems of significant symbols – man's [sic.] behaviour would be virtually ungovernable, a mere chaos of pointless acts and exploding emotions, his [sic.] experience virtually shapeless. Culture, the accumulated totality of such patterns, is not just an ornament of human experience but – the principle basis of its specificity – an essential condition for it.
>
> (p. 46)

The view that culture is added onto some sort of more primordial experience is what Geertz is attacking. He is pointing out how deeply ontologically dependent on culture humans are, and this theme is not sufficiently grasped in CSR and cross-cultural psychology.

He argued that humans are guided by cultural templates more than genetic templates to such a degree that culture completes us constitutionally. Personhood and individuality emerge in and through culture and so drawing a distinction between 'natural' basic bio-chemical strata and more abstract cultural strata is not possible. In other words, the distinction between the enculturated and non-enculturated person does not exist. This approach replaces the stratographic view with what he calls a synthetic view where material aspects of persons (like the nervous system) are not more basic levels of humanity.

This view raises tensions because Geertz (1973) enables a view where the concept of culture leads to a potential decline in the ability to make declarations about human universals. The argument is that people unmodified – in the strongest sense of the term – by times and places do not exist and this claim can be illustrated through his discussion of thought. He points out that intellectual activities involve manipulation of cultural symbols and meanings. Even unconscious thoughts are enabled through interaction with others as children grow and develop. His point is that cultural resources are a constitutive ingredient to human cognition, which means that thinking is not a private process. Other perspectives, such as those we see in Kessing, CSR, and cross-cultural psychology, draw a line between basic universal cognitive processes and cultural information that is subsequently represented in the mind. Geertz points out that it is not so easy to draw such a line, and he argues that supposed universal cognitive mechanisms don't work in humans. Humans, as higher animals, shape the structure of the "input" thought cultural resources and the "output" is not a universal given. Such higher animals are constitutionally shaped through enculturation to such a degree that an a-cultural universal mechanism becomes hard to identify. He points out that the cultural variability can lead to claims about universals that are non-substantive, or banal, due to how general they have to be. The realities that they attempt to address can be so vague that they are not helpful for pragmatically understanding humans. As such, he raises challenges that can lead to a fear of relativism, and this can cause tensions with researchers who are interested in discovering universals.

Bringing the forging together highlights how studying religion involves considering meanings that in a way includes the body. Detailed consideration of such communal meanings is important instead of approaching culture as a synonym for nation. The view of culture I seek to espouse includes the body (see Chapters 3 and 4), but recognizes the strong way that humans are culturally constituted. The issue of universals is therein raised, but James' pragmatism is an ideal resource in such an epistemological context. It in this broad context that I move forward with cultural psychology.

Cultural psychology and studying what belief is about

When we start to consider what a belief is about, we are really addressing the meanings encapsulated in culture. I seek to look at such meanings in everyday life. In the case of the example above, Participant 5's belief that prayer is a way of getting into God's presence amounts to what this belief means. Cultural psychology[1] entails an approach that is interested in what she says and the meaning of her comments on prayer. Jacob Belzin (1997, 1999, 2005, 2010a, 2010b), for example, is a cultural psychologist who has written a substantial amount on religious belief (see also Hampson, 2005). He makes a point that a culturally informed approach is ideal because it has been explicitly concerned with meaning in relation to culture (e.g., 2010b, p. 25). Cultural psychologists, such as Belzin, are interested in how psychological phenomena, like belief, are relational phenomena. They are interested in the way that something like belief is shaped through culture (Benson, 1991; Bruner, 1990; Shweder, 1991). Where psychological phenomena can been defined in terms of mental mechanisms, considering how they are relational phenomena means that they can also be defined through what they relate to (e.g., Belzin, 2010b, p. 28).

Consider how culturally shared stories outside of a single individual provide a repertoire of concepts and plots that can shape belief. Participant 5 is a good example insofar as she unfolds a plot for Blake. It begins with a description of a problem: the impending death of her cousin over which she has little power. She then describes her action as an attempt to obtain healing but offers a narrative twist at the end, and this is the moral of the story. The moral is a statement of belief that prayer is healing insofar as it brings one into the presence of God. The story explains and outlines her belief in terms of where it came from in the form of a common plot structure. Her belief is structured through this plot, and it is structured through normative expectations. She describes begging instead of demanding and so illustrates a communally shaped expectation of the language style in which one should pray (e.g., praying to "Our Father" as opposed to "dear dude"). The point is that such stories come from a socio-communal milieu that makes beliefs sensible and understandable in everyday life. The belief of being healed is situated within a cultural narrative that makes the belief sensible – one can sensibly explain belief on the basis of communal standards – and therein shapes what it becomes.

Cultural psychology is important because of how it addresses such ideas to show the socialization/enculturation of belief. A cultural approach highlights how interpreting belief as an abstract and disembodied phenomenon is misguided (e.g., Belzin, 2010b, pp. 24–25; Hampson, 2005, p. 269). Religious belief should be studied as a relational phenomenon because, at some point, it needs to be articulated in language and so immediately falls into stories that are culturally shaped. Consider as an example how a discussion of an ineffable experience in prayer is itself a cultural narrative. The very ineffable quality is talked about in a normative way that constitutes it. The "ineffable" has meaning precisely because there is a narrative for it. A culturally informed approach is that religious belief is culturally construed precisely because humans cannot do otherwise.

Such an approach to the psychology of religion is one that does not treat the meaning of religious belief as secondary. In addition to resonating with the foregoing on culture, it also fits with James insofar as he acknowledges how belief is discursive and so is shaped in communities (e.g., 1981/1907, pp. 83–84, 96–97). James' approach treats meaning as central and ties it to the community in which a religious practitioner engages. Considering stories and language enables researchers to get closer to belief as it shows up in life. In contrast, for example, a hyperactive agent detection device is normally treated as a blind algorithm that takes stimuli, processes whether or not the stimuli belong to an agent, and then generalizes it more broadly to a metaphysical level (Boyer, 2001; Keleman, 2004). The meaningful qualities of God being a healer and being in God's presence that we see in Participant 5 are not addressed, but such qualities and what they mean to her is precisely what matters if we are to understand belief as it fits into life. Who God is and what qualities are tied to prayer are critically important. To understand our research participants, we turn to the stories about metaphysical agents to address how people actively make sense of the world (Donald, 2001). We would then be able to grasp how a religious practitioner's approach to the meaning of a divine super-agent shapes belief and understands herself in this world. For example, we can apprehend the meaning of the belief that God's presence is healing and see how it organizes her self-understanding when praying for a physical healing that may never come. Precisely such morals are what fit into a practitioner's life and not some vague super-agent accidentally produced.

Cultural psychology provides an approach that enables a discussion of communal meanings, but a problem is that a researcher could simply replace the focus from naturalist mechanisms with a focus on stories and constructions. Religious meaning and *experience* is deeply personal instead of being "nothing but –" a socially constructed story. Remember that belief is not abstract but it is also embodied, and so we must also address the deeply experiential feature of religious belief and in some way, an experiential sympathy on the part of researchers. To be fair, some cultural psychologists like Belzin have attempted to develop the notion of meaning in relation to experience (e.g., Belzin, 1999). His work on the Christian Bevindelijken tradition emphasized how the experience of one's own experiential sureness of salvation is deeply tied to socially constituted narratives (e.g., 2010b, p. 147). To derive his account of meaning, Belzin (2010b) relied upon theorists such as Hermans and Kempen (1993), who often place an importance on the role of language and its ability to construct the world as we know it. This approach to language is good insofar as it enables a discussion of meaning. It has, however, been criticized for neglecting embodiment (e.g., Baerveldt & Voestermans, 2005; Benson, 1991; Cresswell & Baerveldt, 2011; Smith, 2010). That is, we could run the risk of missing what one author named "deep enculturation" as a way of noting how the culture is a deep part of the body that constitutes ostensive realities (Donald, 2001). My goal is to enable a pragmatic discussion of meaningful embodied experience, as opposed to just meaning, in order to really get at truth as something that feels right deep down (see Cromby, 2015). A claim about belief that is disjointed from such experiential aspects of life would not be true and satisfying. Scientists must thereby pay attention to such notions.

Cultural psychology and understanding experience

Mikhail Bakhtin is a philosopher who has inspired work in cultural psychology (e.g., Hermans & Kempen, 1993; Shotter & Billig, 1998). He offers a view of culturally shaped stories that rests upon a discussion of language that can help us understand experience. There are three aspects of his discussion that can bring us to an approach to belief that fits with the life of believers: (1) lived experience, (2) how such experience is socio-linguistically shaped, and (3) how it necessitates a constant sense that there must be someone else to talk with (i.e., otherness).[2]

Lived experience

Bakhtin's early work described how people are caught up in just living life. That is, he described how people act in a flow of experience: "all that which is given, present-on-hand, already realized and available – recedes, as such, into the background of the action-performing consciousness" (1990, p. 43; see also p. 85). For example, he wrote about how we simply shift to the side when we are walking toward someone (ibid., p. 440). This shift is an action that is something we don't consciously attend to because we just do it as a matter of course. When he used the phrase "act-performing consciousness," he was pointing out how people simply think in embodied action; we just do most of life and don't pay too much attention to our doings.

This claim about how humans live their lives has roots in phenomenology (Bakhtin, 1993, p. 6), which is an approach that also concerns itself with how life is experienced (Cerbone, 2006). While Bakhtin is not a phenomenologist proper, he does resonate with the phenomenologically oriented notion that the world seems to show up to people in life. Consider Bakhtin's comment that "[t]he act performed proceeds in an objectively valid context: in the world of narrowly practical ends (ends relative to living one's daily life), in the world of social and political values, the world of cognitive validities (the act of cognition) . . ." (1990, pp. 138–139). When he writes about an act proceeding in an objectively valid context, he is talking first about how we are caught up in life like I described above. When Bakhtin writes about notions like "an objectively valid context" and "the validity of the object," he was addressing how the world appears to humans. Humans live in a world that is immediately experienced with seemingly objective reality. A world shows up to us in a way that is so real that acting against it is to act against what seems to be objectively obvious reality. A potential collision with another pedestrian, for example, does not involve questionable entities which people could or could not believe in, and it is obvious that we need to sidestep a collision (Cresswell & Baerveldt, 2011). People are compelled to believe that the world is the way that they experience it and to do otherwise would seem ludicrous (Bakhtin, 1993, p. 38).

It is easy to see what Bakhtin is writing about in terms of pedestrians that one cannot walk through, and these sorts of things are what he meant with the phrase "world of narrowly practical ends." Practical ends like pedestrians are not what Bakhtin is primarily concerned with, and I follow suit because practical ends do not tell us a lot about studying religious belief. The "objectively valid context," as I cited above, also involves "social and political values" and "cognitive validities," and these are just as real to people as a pedestrian with whom we are on a collision course. Values, thoughts about the world, and ethical values are also part of the world in a way that seems just as objective. What Bakhtin is telling us is a claim that is strikingly similar to James: beliefs are manifest in the world with seemingly concrete reality because "non-incarnated thought, non-incarnated action, non-incarnated fortuitous life is an empty possibility" (Bakhtin, 1993, p. 43). These sorts of phenomena are also part of the world as it is manifest to humans *in the same way* that the objectivity of a pedestrian is.

Consider the case of Participant 5 and her belief that "God *is* healing." She is making a claim about an objective reality that seems real to her insofar as she is describing how something *is*. Consider how the structure of the explanation of that belief also points to the way that it is objective: "begging for the medicines to work, begging for the community support we needed, and all of that came through in the gentlest most sweet most supportive way possible and the thing I learned from that experience was that God doesn't choose to heal or not, God is healing." The belief is predicated on an answer to prayer being answered in the "most supportive way possible." These are not formulated as descriptions of a potential perspective because they are formulated as statements of clear facts. She is outlining a belief about something that she learned from experience, and so her talk illustrates how it is not simply something locked inside of her head. The value of Bakhtin's ideas is that we can understand how phenomena like belief are experienced as part of the seemingly objective world.

A further way to understand this idea is through Bakhtin's discussion of the "emotional-evaluative tone" (1990, 1993). This notion involved the observation that emotion shows up in life as part of the world. We experience the world *with* emotional-evaluative tones and *not* a place where emotion is laid over top of a neutral reality. After all, "I experience the object of my fear as fearful, the object of my love as loveable, the object of my suffering as oppressive . . . but I do not experience my own fearing, loving, suffering. My lived experience is . . . the *whole* of myself in relation to some object" (Bakhtin, 1990, pp. 112–113; original emphasis). That is, emotion is related with life so when Participant 5 says that community support "came through in the gentlest most . . . sweet most supportive way possible" she is saying that the support *is* gentle and sweet. The emotion is in the action from her perspective. We can also see how she illustrates the way that emotion involves an evaluation. Her description of the actions of others is emotional, and it evaluates the action as gentle and sweet. For humans, the world *is* emotionally evaluative in their experience of it. It is in this manner that people evaluate the world as they affectively experience it, and it is, consequently, as seemingly objective as phenomena like pedestrians.

Belief, of any sort, thereby simply includes such an emotional-evaluative quality. An experience of the world as it shows up in life is the seemingly irreducible objectivity of religious sentiments that are bypassed from a naturalist perspective. Turning to Bakhtin enables us to address such compelling everyday experience that is neglected. The foregoing gives us a way to approach religious belief in terms of experience and addressing what belief is about. Our goal is to look to develop an approach to belief that fits with life. I illustrated how this approach involves a consideration of what belief is about by turning to cultural stories and the language people use if we want to get at truth being something that fits with life. Scientists are required to turn to the use of everyday language and the narratives people tell in order to grasp belief. The most recent discussion addressed lived experience in terms of what it is like for people. Ideas and emotions are things that seem objective, but I haven't really tied such lived experience to the world outside an individual. Cultural psychology grants insight to meaning because it allows a discussion of what belief is about and so goes beyond an underlying concern with naturalist computational laws. The challenge at this point is to put together culturally shaped meaning and lived experience.

Socio-linguistically shaped lived experience

It is possible to rise to this challenge by continuing with Bakhtin and explicating some work that I have done previously on him and language (Cresswell, 2011; Cresswell & Baerveldt, 2011; Cresswell & Teucher, 2011). Bakhtin writes about how language is an interconnected web of meanings generically used by a community and in so doing, refers to speech genres (Bakhtin, 1986, pp. 76–78). A community of religious people, for example, would have a speech genre that includes a particular repertoire of terms. Participant 5 uses phrases when she talks about genuinely praying with "an open heart" and praying "in a very solid way" where she "begged" for someone's life. These are phrases, of course, that other people use, but they are generically used within her community more often than not. Such generic phrases that mark a Christian approach that is accompanied with a particular embodied stylistic way of expressing it. She talks about being flat on her back, and the tone of her expression in the recording of the interview is slow and emphatic.

There are two implications that are inspired by this notion of speech genres. The first is that they involve a communally shared shaping of the world (see Cresswell, 2011, 2014). A speech genre involves a manner of apprehending the world that is shaped by a community. That is, the narratives in a speech genre shape the way that people understand the world; they shape meaning by way of shaping how people apprehend the world around and within them. Participant 5 is explaining a belief about things like getting healed and how the presence of God is healing. This discussion is tied to a community that would say such things and understand such concepts in a way that makes the experience of not seeing the healing of her cousin sensible. The lack of healing is understood through the stories and terms that shape how she understands it.

A second implication is that speech genres are embodied and so include experience. Consider how Bakhtin argued for in intimate relationship between language to embodied notions such as an emotional-volitional tone (1990, p. 31). This coupling of language with experience comes from participation in life with others. In childhood, humans are socialized into how to feel about the world (Bakhtin, 1990, pp. 50–52, 153–154; see Chapter 3). For example, the actions of religious caregivers could socialize a Participant 5 into experiencing the situation of her cousin's cancer as "shocking," and so it makes sense for her to say that "we all sorta went into shock." This is a term that she has been socialized as part of an experience. Consider the experience of "presence" as another term into which she has been socialized. It is a piece of linguistic jargon, but it is bound to embodied experience of some sort. It is something into which one tries to get and so the term means that there is a something out there to experience, and prayer is what gets one there. Through praying and talking with others, she was part of subtle socialization that amounts to an acquisition of an emotional-evaluative disposition as to what "presence" experientially is.

Hence, one can see how the living experience of the world is constituted by communities in which we are participants. A significant part of language is the experiential insofar as generic terms used to describe an experience are not separate from the experience. For example, learning to pray is learning to express stories and terms like "presence." The result is an idea inspired by Bakhtin that seeming objectively lived experiences are socio-linguistically shaped. To grasp truths that fit with a community, one must learn the speech genre of a community that includes its experiential depth.

An important implication for my purposes is that belief involves experience shaped through communities. A repercussion is that language is bound to experiences socialized in a community through language and researchers who learn such stories and language learn the concomitant experience. Doing so enables a better fit with what participants live. This approach further differs from the naturalist approach because it deals with what beliefs are about in addition to experience. A researcher informed by a cultural approach is free to apprehend the lived experience of a community and does not have to abstract herself from experienced meaning through pursuing objective methods. That is, when trying to apprehend the stories involved in payer, a researcher apprehends the emotional-volitional tones that are part of them. Bakhtin inspires an inductive approach that can potentially address the deeply meaningful *and* seemingly immediate experiences people have as part of their participation in religious communities. Here, we can get at religious belief and do so in a way that what we find can fit with those we study.

Constant experiential otherness and religious experience

At this point, we have an approach that can help us look at belief in a way that fits with the experience of believers. Participant 5's discussion of "presence" is predicated on there being a God that transcends her and into whose presence she seeks to be. Included in this discussion is an experiential necessity of there being an other that is beyond her. I do not take the position that only believers could research religious belief because a robust research program could be one that allows an atheist to understand a believer. There is potential for a cultural psychology inspired by Bakhtin that leads to a sympathetic position on the part of researchers. It inspires understanding of the experiential necessity of transcendence ubiquitous across religions and spirituality (see Slife and Reber's (2009) claim that theism necessitates a position presuming a supernatural other).

Among academics who work on Bakhtin and cultural psychology, there is a well-known phrase attributed to him: "there is no alibi in being" (1993). An alibi is an excuse or way to get out of something, and what Bakhtin meant by this phrase is that people are never free from having to answer for themselves. Being a human means that we constantly have to account for ourselves. In the case of Participant 5, it is pretty easy to see how she illustrates this claim. She is accounting to Blake for her belief that prayer is a way to get into God's presence. Her situation is one where a young man is asking her about religious doubt and her beliefs specifically. In life, however, we are always doing the same in very subtle ways. "No alibi in being" articulates how life with others entails people constantly accounting for themselves. Even though people just do life without too much reflection, a lot of what we do and say is directed to another. For example, even in Participant 5's prayer, there is an orientation to a community of people that expect people to pray for others who have cancer. In the background of the prayer is a prior expectation by her and her community that she ought to be doing so. Because our way of understanding and thinking about the world is steeped in the speech genres that we learn, every thought is a response to a community. The meaning of action and experience is always relative to another from the past, the present, or the anticipated future. In the research community, we refer to this notion as "otherness" because we are always oriented toward another that transcends us.

One of the reasons I advocate a turn to Bakhtin is that he was particularly interested in religious experience in reference to such seemingly transcendent otherness (e.g., 1990, p. 47) and specifically did so in reference to the act of prayer (ibid., pp. 138–150). Bakhtin noted that people can become misguided when they consider the notion of confessional prayer in particular, and the discussion of this problem can influence how we think of transcendence in general: such activity is commonly thought of as a private act (e.g., ibid., p. 138 & 141). Usually, Bakhtin argued, normal interactions are ones where we depend upon another for closure so that we can finish something off. Consider this section from the conversation about prayer.

Turn	*Speaker*	
3	Participant 5	in his presence you heal
4	Blake	Yeah
5	Participant 5	so the trick was to get into his presence
6	Blake	mmm, and prayer was
7	Participant 5	and prayer was the tool for that.

In turn three, Participant 5 finishes her sentence with a claim that in God's presence one heals, and she uses the second person pronoun ("you"). Blake responds with "yeah," and this seemingly insignificant conversation filler is very important because it closes off the claim by validating it. If Blake had said "mmm," she would have continued to explicate the claim. The agreement enacted with "yeah" concludes the claim and lets her transition to something else in turn 5. We often depend on others to conclude our thoughts and expressions, and these are constructed in the moment (see Gergen, 1999b).

As I argued elsewhere, we also rely on others for us to get a sense of our own selves (Cresswell, 2011). When life is generally lived in a constant flow, we are just doing what feels normal. We can only escape this openness of life when we see ourselves through the eyes of the other (ibid., p. 86). Through another's eyes, people become objectified as an instance of a speech genre. Participant 5 probably lived her life and conducted her prayer because it just felt like the next right thing to do. It was when she had to account for herself to Blake that she had to think through another perspective on herself. She is now able to see herself through another person's eyes instead of just doing the next right thing. In everyday conversation, one may be disquieted, and seeing through the eyes of another allows one to apprehend how such disquiet comes from values in one's community. For example, Participant 5 builds a case explicating why she has the belief that she has and so can see herself as someone who does something that others in her community do.

Bakhtin noted that, in confession, it seems like one cannot get outside of just living because there seems to be no present other. Confession is potentially meaningless without an addressee and closure insofar as it seems to not be connected to the otherness. Always living in an experience of otherness means that human action cannot exist without it. In confession, Bakhtin claimed, there is a necessary experience of God as the bounding limit to confessional self-accounting. Humans find *approximate closure* because it is simply not experientially possible to live as if one is alone, and I propose that this notion can be generalized to transcendence as well. There always has been someone relative to whom we take a position and someone who is interested, even when not present. According to Bakhtin, the deeper the solitude, the deeper the necessary reference to God:

> Pure, solitary self-accounting is impossible; the nearer a self-accounting comes to this ultimate limit, the clearer becomes to other ultimate limit . . . the deeper the solitude . . . with oneself, and, consequently, the deeper the

> repentance and passing-beyond-oneself, the clearer and more essential is one's referredness to God. . . . Outside God, outside the bounds of trust in absolute otherness, self-consciousness and self-utterance are impossible, and they are impossible not because they would be senseless practically, but because trust in God is an imminent constitutive moment of pure self-consciousness and self-expression.
>
> (1990, p. 144)

One cannot obtain momentary closure on the self but is nevertheless driven to find it in confessional prayer by virtue of how we have always been doing so. In the case of transcendence in general, being unable to find closure on the self is the transcendent experience of seeking the otherness that we have always lived in.

Moving forward

Such an approach brings us to a place where we can look at what belief is about and the experience it entails and specifically inspires a way to further sympathetically apprehend religious experience. It gives us socio-linguistically shaped meaning and the experiential necessity of transcendence that is often central to religious experience. Whether a believer or not, the culturally informed psychologist comes to the same place: *a sympathetic understanding of the experiential necessity of otherness*. Relative to CSR inspired by naturalism, what I propose is practically advantageous. It addresses meaning and experience as studying communally shaped beliefs involves cultivating experiential sense of them.

In this chapter, I have tried to offer an approach to religious belief that is often bypassed in mainstream psychology of religion and CSR, in particular. What I have proposed enables research to fit with the experience of believers because a discussion of culturally shaped meaning and the experiential aspect of language enables a discussion of experience. It is based on our experience with the world and others that we necessarily orient to another – albeit a rather nebulous other. I hope, thereby, that it is possible to fall in line with James' vision and to work towards a cultural psychology of belief that does not reduce it to a "nothing but –." The approach I provide is not exhaustive, however. There are limitations, and work remains to be done. One limitation is that of developing an understanding of how a pragmatic cultural psychology of religion could relate to current work in the CSR. The foregoing does not deal directly with CSR, and more needs to be done in order to enhance our work therein. There has been substantial work in CSR, and I am not suggesting that everything prior to my proposal be thrown out. Rather, for example, work needs to be done to build bridges to CSR. Recall that another criterion that establishes truth is generativity. It builds on fit insofar as fit with life is not just a static fit with other aspects of experience at one moment of time. A true scientific belief, James (1981/1907, 2011/1909) argues, is one that fits the dynamics of life to do more. For a belief to remain true, it must generate something for us was we move forward such that it makes future activities smoother. Truth is "essentially bound up with the way in which one moment of our experience may lead us towards other moments which it will be worthwhile to have been lead to" (1981, p. 95). So far, CSR researchers have been moving forward together insofar as their work has generated lots of ideas and discoveries, but they have not moved forward in light of the ideas cultural psychology opens up for us. The problem is that it has not been generative for religious believers due to abstract and disembodied naturalism. The goal with this book is to broaden the way we approach the study of religious belief so that we can move forward in a broader range of circles. An approach to truth that can fit for religious believers by looking at belief as concrete and embodied is what is needed to move us along.

The challenge lies in creating a point of interface between the strong cultural approach outlined above and CSR because such an interface must be truthful to more than one community to move us along. A fruitful place to go would be one that relates the above to newer cognitive theory. Consider recent work in embodied cognition that draws on phenomenology and gives culture a substantial role (e.g., Thompson, 2007). A viable bridge between a cultural psychology of religion may be possible by pushing theory in the psychology of religion toward such new work, which is the subject of the next chapter. The next two chapters explore a unique approach to the development and evolution of human cognition with the aim of providing an approach to cognition that better includes the cultural quality of belief outlined in here. My agenda is to offer an approach to research that fits with the lives of believers while also fitting with the community of researchers (Chapter 3) and building a generative bridge between CSR and religious believers (Chapter 4).

Notes

1 This moniker is a debated one. I am promoting a strong form of cultural psychology that is reflective of the *Oxford Handbook of Culture and Psychology* edited by Valsiner (2012). This is a strong form of cultural psychology that stands in contrast to reincarnated cross-cultural psychology as exemplified by Kitayama and Cohen (2007).

2 I published an argument similar to this in Cresswell (2014)

Chapter 3

Don't throw the baby out with the bathwater I

Development of religious belief[1]

Abstract

Previous chapters demonstrated that the phenomenon of religious belief is quite different from the kind of phenomena the cognitive science of religion is studying. To fully understand belief, researchers must apprehend socially constituted experiential realities that communities of believers embody in their languages and cultures. This chapter and the next turn to a discussion about how taking the constitutive force of language and culture seriously does not mean the suspension of good science. We can still have good scientific practice that fits with both religious and scientific communities, which is a criterion for truth for James. Specifically, this chapter addresses how cognitive science of religion could be enhanced if provided with the theoretical articulacy to grasp the constitutive role of language in the formation of religious cognition. Religious ritual and the everyday ritualistic use of language is discussed to specify what CSR stands to gain by taking the foregoing seriously. The proposal herein draws on dynamic systems theory, which offers an approach to cognition that enables the study of religious belief as it emerges in life and offers a theoretical framework that allows researchers in the cognitive science of religion to proceed in harmony with cultural psychology.

Introduction

Inspired by James (1981, 2011), I introduced a second criterion for truth as it fits with a community. To get away from the problems associated with abstract and disembodied truth, it could look like objectivity is lost. This implication is correct, but the inference that we collapse into subjective fancy is not, and understanding how this is so is what proponents in the community of cognitive science of religion (CSR) need if they are going to move forward. James (1981, 2011) repeatedly argued that truth is not a matter of the mere proclivities of subjectivity by referencing standards of fit as socio-normative in quality. For James, the truth of belief is deeply entwined with living with others with whom we agree on what counts as true. The truth about naturalist mechanisms in CSR works for that community, and it is time to couple this to a broader approach that speaks to a community beyond its own. That is, the foregoing argued for a culturally informed approach so that we can understand the beliefs but we cannot, however, just abandon CSR and glibly dismiss the interest in understanding human cognition.

To show how it is still possible to have a theory of CSR that can account for religious believers in a way outlined above and still matter for the community of researchers in CSR, I will discuss an elegantly simple study conducted by Keleman and Rosset (2009) and work through this study as an illustration in this chapter. They start from an interesting observation that people tend to be resistant to scientific explanations for the world. I subtly broached this topic in earlier chapters when I discussed how religious believers often ignore the claims of cognitive science. Where I argued that this phenomenon happens because CSR bypasses what belief is about, Keleman and Rosset pick up a slightly different aspect of this discussion. They argue that people don't pay attention to science, in general, because they have hard-wired conceptual biases that inhibit the ability to represent the world in the way that scientists represent it. That is, they argue that humans have an intuitive inclination to process information about the world according to a principle of "promiscuous teleology," which is taken to be an automatic tendency to explain phenomena according to a purpose. For example, "[w]hen asked about properties of natural entities like pointy rocks, children prefer teleological explanations over physical – causal ones, endorsing that rocks are pointy 'so that animals won't sit on them', not because 'bits of stuff piled up over time'" (p. 138). The goal of Keleman and Rosset was to show that, while it may seem that such explanatory errors are committed by children, all humans tend to impute cause to many natural living and nonliving things. Promiscuous teleology may not be a passing stage of immaturity, they argue, because it is possible to make this intuitive response manifest in adults when there are significant demands placed on the ability to process information. As such, the objective of Keleman and Rosset's article was to demonstrate that educated "adults should display scientifically unwarranted promiscuous teleological intuitions when their capacity to inhibit more primary purpose-based intuitions is impaired by processing demands" (p. 139).

Keleman and Rosset (2009) concluded that adults have a suppressed tendency to explain living and non-living phenomena by reference to purpose. This predilection involves treating nature as a goal-directed self-preserving organism and so highlights how religious belief could be predicated on this seemingly natural tendency. The authors stated that "it remains for future research to clarify what factors and beliefs promote and underlie college-educated adults' promiscuous teleological ideas, as well as the relationship between children's and adults' teleological beliefs" (p. 143). Discussing such factors is exactly what can reveal an approach to cognition that allows for a discussion of meaning while still offering theory about cognition that is acceptable to people in CSR. The goal of this chapter is to respond to Keleman and Rosset's challenge in a way that pushes CSR in new directions.

Consider how researchers in the CSR have propagated an interpretation of phenomena like promiscuous teleology as existing because they rest upon universal pre-given cognitive architecture (e.g., Barrett, 2007; Boyer, 2001; Cohen et al., 2008; Guthrie, 2002; McCauley, 2011; McCauley & Lawson, 2002; Spelke & Kinzler, 2007). That is, humans are supposedly endowed with many domain-specific modules that enable, for example, promiscuous teleology, theory of mind (e.g., Bering, 2006, 2011), agency detection (e.g., Barrett, 2004; Barrett et al., 2001), and the retention of religious ritual (e.g., McCauley & Lawson, 2002; Pyysiäinen, 2002). Such modules are taken to conjunctively work to produce religious phenomena, which in turn lead to religious belief. Promiscuous teleology is taken to be emergent from hard-wired mechanisms. Much of the work in this area involves examining children with the hopes of assessing them at a young enough age to preclude the meaningful influence of enculturation (e.g., Barrett, 2012; Keleman, 2004; Spelke et al., 1992; Spelke & Kinzler, 2007). That is, it doesn't involve examining the meaning of teleological thoughts as something that matters per se. Instead, the content of what might be meant by such teleology is not studied in its own right – as I argued in Chapter 2 – because mechanisms shaping the phenomena are of interest.

If someone were to examine Keleman and Rosset's (2009) study, one may say that it is strange to talk about an expression of religious belief in adults and stranger for me to write about it in a chapter on development. Like Keleman and Rosset, I want to write about the expression of belief in adults because this is what CSR is trying to ultimately explain. I have shown in previous chapters that the phenomenon of religious belief is quite different from the kind of phenomena CSR is studying. To fully understand belief, researchers must apprehend socially constituted experiential realities that communities of believers live as adults and not just children (Hampson, 2005; Reber, 2006; Slife & Reber, 2009, 2012). Language and culture, I argued above, shape the realities people believe in and experience. The culturally specific beliefs entwined with promiscuous teleology is what I propose can be studied and doing so involves turning to look at how it was/is constituted in language practices. CSR bypasses such lived realities by focusing on individualist mechanistic processing and a more inclusive CSR is possible, if not desirable.

This chapter addresses how we can study the cultivation of cognition in infancy to show the development essential for belief and the meaningful experience it entails later in adulthood. It does so by addressing the presupposition that enculturation does not happen until a child is well into the use of language. CSR is predicated on the notion that innate mechanisms form the basis of cognitive architecture whose operation precedes the acquisition of language and concomitant enculturation (e.g., Sperber, 1994, 1996). However, language is a ritualistic practice that is integral to psychological development from birth, if not earlier (see Cole, 2003; Fogel, 1993; Wertsch, 1985). These challenges indicate that CSR could be enhanced if provided with the theoretical articulacy to grasp the constitutive role of language in the formation of religious cognition. In so doing, it is possible to inform CSR with a theory of cognition that considers the meanings entwined with promiscuous teleology.

The objective of this chapter is to offer a theoretical approach to the ontogenetic development of religious cognition that considers the role of ritualistic language practices in the formation of religious cognition and offers a view of cognition that is compatible with this consideration. This chapter attempts to accomplish this objective by first outlining a sociocultural approach to the development of religious cognition. Drawing on the ideas introduced in Chapter 2 enables an articulation of how religious cognition is cultivated in early childhood. Central to this extension of Chapter 2 is the role of the regular ritualistic use of language that shapes cognition. I then specify what CSR stands to gain by taking the foregoing seriously before turning to dynamic systems theory, which offers an approach to cognition that naturally incorporates a sociocultural approach (see Elman et al., 1998; Noë, 2009; Prinz, 2009; Thelen & Smith, 1996). Such work should lead to theory that enables the study of religious belief as it emerges in life and provide a theoretical framework that allows researchers in CSR to proceed in harmony with cultural psychology.

Culture, language rituals, and the development of religious cognition

Cognition is generally understood to be about thinking. It includes higher-order thoughts such as knowing what one's religious beliefs are (Harré, 2002; Thagard, 2005). It also includes a discussion of the kinds of tasks or activities of which we are not always aware. As such, cognition is about thinking in a broad sense that encompasses activities ranging from reflective thought to dynamic embodied dispositional behavior; the latter is captured in Keleman and Rosset's work on promiscuous teleology while the former is bypassed.

Studying religious cognition is a difficult task, as we cannot simply peer into the heads of others and physically see their religious beliefs. Adding to this complication is the way that cognition involves meaningful symbols, which are notoriously difficult to pin down in a scientific manner (Harré, 2002). Generally, cognitive scientists deal with symbols through a discussion of mental representations: "To address the crucial questions about the nature of mind, the psychological experiments need to be made interpretable within a theoretical framework that postulates mental representations and procedures" (Thagard, 2005, p. 8). Cognition is often about symbolizing the world around us inside the mind, and these symbols are representations (Boden, 2006). Consequently, "[h]ere is the central hypothesis of cognitive science: Thinking can best be understood in terms of representational structures in the mind and computational procedures that operate on those structures" (Thagard, 2005, p. 10).

Consider how Keleman and Rosset (2009) demonstrate promiscuous teleology in adults. They showed participants sentences that the participants had to judge as "good" or "bad," which means judging if a given sentence is incorrect or correct. A correct statement would be one that is scientifically correct where an incorrect one would be one that involves promiscuous teleology (e.g., the sun shines for plants to grow). Half of the participants were put under strain insofar as they were required to decide very quickly if a given sentence was "good" or "bad." The other half were not put under time pressure. Keleman and Rosset's logic was that requiring participants to answer quickly would make intuitive mental processes appear because participants could not critically evaluate their answers. Participants would be forced to fall back on available computational structures. Such intuitive mental processes would be the promiscuous teleology that we see in children, and so the research was designed to reveal that people don't grow out of the computational procedures that we come with at birth. Keleman and Rosset found that participants who had to answer quickly were indeed more likely to endorse scientifically unwarranted sentences as being "good" and so showed promiscuous teleology relative to those that could critically evaluate their answers. This effect held even when the researchers eliminated potential extraneous influences like personal beliefs and scientific knowledge to which participants had been exposed previously. As such, it would seem to be an intuitive mental computational program is running.

This approach misses the importance of language and the reflective side of thinking as it entwines with intuitive promiscuous teleology. Language acts, such as the constitution of meaning of teleologically promiscuous sentences, takes second place to underlying mechanisms that produce the behavior. Harré (2002) notes, however, that symbols are central to cognition, yet it may be premature to presume that mental representations capture this centrality. He argues that cognitive tasks include a central role for language as the most definitive form of symbols. Relying on inner representations may bypass the way that Black's religious cognition is deeply entwined with the repeated sociocultural experiences of language learning.

Harré's (2002) claim is highly resonant with cultural psychology, which also purports that the symbols we learn in repeatedly practicing language are deeply tied to our thinking, and this includes our dispositional activities. As I mentioned above, cultural psychology is predicated on the notion that cognition is constitutionally entwined with culture (see Valinser, 2012). A cultural approach often turns to language as a primary consideration when establishing a case for culturally constituted cognition (see Burkitt, 2013; Cresswell & Teucher, 2011). Participating in language, for example, entails narratives that shape the concepts upon which cognition depends. The sentences that assess promiscuous teleology must thereby be understood as more than simply the use of lexical communication used to express thought because it constitutes his religious cognition. As such, any type of religious cognition and what such belief entails is shaped in language.

Furthermore, language is deeply entwined with the body (Burkitt, 2013; Cresswell & Baerveldt, 2011; Cresswell & Teucher, 2011; Cromby, 2015). Bakhtin (1986) discussed the notion of speech genres to articulate how there are generic ways that communities linguistically embody what one ought to do and say. He pointed out that, without fully articulating an ideology through lexicon, generic conventions are tacitly enacted and continually re-established by members of a community. A member of a community comes to express him or herself in the generic style of the community and feels compelled to express that style.

Therefore religious belief and the concomitant experience entailed in such cognition are constituted through, and bound up in, language and culture. To highlight how CSR could be improved, I will turn to the ontogenetic development of religious belief and how cognition is shaped through language practices in childhood. Just like CSR explains adult religious cognition by looking back to development, I will do the same using promiscuous teleology expression as a window to do so.

Researchers are advancing sophisticated theories about how participation in dynamic interactions shapes an infant's cognition (e.g., Cole, 2006; Fogel, 1993). Central to the cultivation into the embodied practice of language is "co-regulation" (see Fogel, 1993). Co-regulated interactions between a caregiver and infant unfold in a dialogical back-and-forth interchange of trial and error where the actions of one party are entwined with the actions and reactions of another. Infants engage in a rhythmic flow of dialogue turn-taking with vocalizations, such as babbling and cooing (Hsu & Fogel, 2002). Caregivers reciprocate with infants' vocalizations with affirming babble/coo versions of their own and pause to encourage responses.

It is important to maintain sight of how caregivers' actions are enactments and reenactments of generic language practices within a community that tacitly enculturates infants' cognition (see Burkitt, 2013; Cresswell & Teucher, 2011). Although these interchanges take place with infants expressing a pattern of sounds, rather than fully formed words, it does not mean that they are not participating in language practices. Since these language practices are embodied, infants not only begin to differentiate among words and sounds but they begin to *feel* differences in tandem with them. Greenspan and Shanker (2007), for example, demonstrate how infants express elementary global states that correspond with physiologic nervous system arousal. An infant cries out loud, and the emotion is totalizing, as if its entire being is brought to bear in the cry. An infant's affect involves global sensations of comfort or discomfort, pleasure or pain, and so forth. During co-regulated interactions, emotion is produced and subtly transformed or repeated interactions. When a caregiver regulates an interaction with excited faces, gestures, and affirming sounds, an infant will begin to tailor the global emotion into a more refined repertoire of emotions. Such co-regulation enables infants to follow along with a caregiver's more sophisticated emotionality into more sophisticated emotional expressions.

It is crucial to remember that infants not only participate in the emotion of caregivers, but that the language through which they use to participate in the emotion of caregivers also entails the language of religious belief. Participating in the word "no" can provide an example of how co-regulation works. A caregiver verbalizes the word "no" to an infant who reaches for a dangerous or off-limits object. Significantly, "no" is also physically enacted. The concept of "no" becomes a bodily experience by a caregiver removing the object from an infant's reach or removing an infant from the area in which the object resides. Through continual interactions, a pattern will emerge around the use of "no" as the concept begins to enact various types of restriction. The experience is also bodily as far as a caregiver accompanies "no" with emotional expression. When a caregiver uses "no," I am arguing that it is not employed in a vacuum as it is expressed in conjunction with other words and actions. For example, "No, we shouldn't eat that" with an act of taking something from a four-month-old's mouth is a common kind of action involving embodied language. This investment of emotion in conceptualization and categorization in language tacitly entwines bodily feelings with the development of language. How does this developmental process connect with religious belief more specifically?

To grasp how the religious belief connects with such developmental actions, I will explore the notion of ritual with a little more care. From a cultural perspective, ritual is much broader and cannot be separated into different kinds so easily as CSR, which demarcates religious rituals from other sorts of recursive action (see below). The reason that it is hard to draw a distinction is that what researchers in CSR would call religious rituals are integrally integrated with other daily rituals, which cultural researchers generally describe as "cultural practices" (e.g., Cole, 2003, p. 188). Cole defines ". . . cultural practices as activities for which there are normative expectations for repeated or customary action" (2003, p. 188). They are the rhythmic interactions that proceed in a consistent and orderly manner. For example, Saxe (1996) shows how small interactions between caregivers and children proceed in a dialogic manner where conversation turn-taking is regular and predictable. A caregiver offers a question, followed by a response from a child, and then a follow-up response from a caregiver. The regular rhythm of the interaction was predictable, and Saxe highlights the small interactional regularities that tacitly enculturate children. These were cultural practices as far as they were normatively structured, but they were also dynamic. Mothers, for example, adjusted their responses with a correction, affirmation, question, and they did so as an adjustment to children's level or performed understanding. When we consider ontogenetic development, it happens largely through such practices that have a unique temporal quality to them. Cultural practices often involve an imagination of the future where another level of performance is imagined by an adult or more advanced peer (see Cole, 2003). Most cultural practices happen in a way that is oriented to a future state – a child understanding counting principles, for example. Parents' imagination of child's future becomes an enacted constraint that enables a child to develop in normatively expected ways. The cultural practices are constant re-enactments of dynamic normative expectations, which means that such recursive patterns of action are rituals by virtue of their directive quality. They are not just repetitive happenings, but the daily rituals that illustrate how culture and cognition are related in people's daily activities (Miller, 1996; Saxe, 1996).

The most central rituals (cultural practices) are the practice of language. Language is meaning construction as far as the speaking shapes what can be known and understood (Miller, 1996). If one has the words for something, it can be understood through articulation. Consider the role of storytelling and how narrative can be used to structure how something is understood. In Miller's (1996) discussion of the structure of narrative practices, she showed how the kind of story that we tell to children shapes what we know. Telling a story about patience and waiting as opposed to one about agency differentially structures how children see their social obligations: to prioritize communion or agency. Caregivers from different cultural practices structure narrative differently, which then structures how we perform our psychology. It is for this reason that language is a form of situated action that can be used to construct what we know and importantly, what we experience about the world. Language, in this sense, is a practice into which children are born and are immersed within as caregivers enact it. It is for this reason that some researchers say that development is orienting to systems of meaning (e.g., Cole, 2003; Miller, 1996). Development is about learning to work within the flow of language that is used in generally predictable (ritualistic) ways. We can understand development when we see how children create meaning in everyday life through developing their own competent practice of language. That is, children begin to mediate the world through linguistic symbols that are presented to them in the ritualistic interactions of normative expectations in the use of language. As a child develops competence in language, we can see how people maintain and reproduce the society that they live in by participating in culturally organized ritualistic practices.

Cultural practices are rituals, and a central ritual of this sort is the use of language, which has profound implications for how the world is shaped. Cole wrote that "there is a fundamental structural change during ontogeny such that an organism that lives in a culture but cannot make use of it becomes one for which mediation of action through culture becomes 'second nature'" (2003, p. 181). The interesting point is that language becomes something that a child learns to enact but it becomes second nature. Consider how Keleman and Rosset (2009) draw on sentences that entail promiscuous teleology, but they also entail an experiential confidence. Cole is observing how the ritual of language can shape what we take to be given about the world, and this can be extended to promiscuous teleology in a way that is much deeper than "ambient cultural religiosity" (Keleman and Rosset, 2009, p. 138). It is in the language we speak, and so it is in the bodies we live. I am not addressing mere proclivities of perspective but about what is taken to be real. The rituals that caregivers engage are activities that constitute a milieu, and so the relationship between text and context is not something that is given but something that is established in negotiation (see Miller, 1996, p. 195). The world is shaped by language as we ritualistically practice it, but there is not a static world in the sense that caregivers and agents are active participatory agents in constituting it. Adults may initiate creation of niches, but the contexts are joint accomplishments and become more so as a child grows into a level of competence enabling more and more agency.

To put flesh on what I am arguing, consider the notion of promiscuous teleology as described by Keleman and Rosset (2009). While these theorists are definitely not cultural psychologists, the phenomenon is illustrative of my point. Promiscuous teleology is teleological in that a child believes a general super-agent causally underlies happenings in the world through purpose and design. It is promiscuous in that a child invokes any kind of non-specific super-agent as the primary cause. As I have elaborated more extensively in the previous chapter, Bakhtin wrote about how language is filled with expressions oriented to others as if they were present even when they are not (Bakhtin, 1990). He articulates how we are always addressing another person in a way that invokes an unseen agent because language is always addressed to another. He points out that language includes what we ought to do considering something like an omnipresent other. This kind of addressivity is implicit in the "we" evoked in the illustration of "no." As "no" is enacted, it is done in a constellation of language so oriented to an ever-present other that humans address. That which CSR proponents call promiscuous teleology could just as easily be the enculturation of cognition in ritualistic practice of language. By the time we see adult language, the orientation to another is so deeply embodied that one "knows" about the existence of such causality with verisimilar objectivity.

Such an approach to the ontogenetic development of religious cognition bypasses a classic binary between cognition and body, demonstrating that how something like promiscuous teleology cannot be teased apart from how people come to *experience* the world through the everyday rituals of language. It is through ritualistic language and cultural practices that an infant acquires religious cognition such as promiscuous teleology. The result is that we cannot think of the meaning of the sentences only in terms of mental representations because they are fundamentally bound to the linguistic and cultural practices in the world. The flow of life with others is central to our cognition, and an adequate account of cognition needs to be careful not to slide into too much of an individualist paradigm.

CSR runs the risk of falling too far into an individualist paradigm, and this risk can be seen in how it approaches the notion of ritual. McCauley and Lawson (2002) write a detailed and elegantly careful book on the topic of religious ritual in CSR. They highlight the central issue that begs for a discussion of ritual, and this issue is the transmission of religious knowledge from one generation to another. Religious rituals are one of the central means by which religious ideas are transmitted from one generation to another, but they argue that one must first grasp what demarcates a religious ritual from other forms of ritual. A religious ritual is recognized by how it involves some sort of representation of a super-agent. The representation of the super-agent and the roles that involve this agent distinguish religious rituals from other recursive happenings. There are other features that demarcate a religious ritual such as there is often a "patient" or recipient of the activity. This role often involves a person receiving something from a super-agent, which can be a conduit (e.g., priest) or direct recipient (e.g., parishioner). McCauley and Lawson outline how such elements are generally present and attempt to do so in a way that can accommodate the substantial variability in religious rituals. The central point is that these elements are part of a recursive activity that passes on religious beliefs from one generation to the next.

The specifics of the how transmission occurs depends upon the kind of ritual that is at stake. Whitehouse (2004) draws on McCauley and Lawson (2002) to outline two general kinds of ritual that enable transmission in different manners.

One kind of ritual is what Whitehouse and McCauley and Lawson call rituals with high "sensory pageantry." Such rituals are ones that have a substantial amount of sensory stimulation and excite emotions. A ritual of this sort tends to have powerful imagery and dramatic activity. They claim that rituals with high pageantry have high emotions, and this leads to emotional provocation in the participants. Provoking emotions makes a religious ritual and the representation of the superagent more likely to be retained. The increased likelihood of retention naturally extends to a higher likelihood of transmission.

Another kind of religious ritual is what Whitehouse (2004) calls "doctrinal/routine" rituals. These kinds of rituals are the more mundane regular rituals that are less emotional and contain less imagistic use of symbols. They are about the regular repeated teachings that are often associated with religious doctrine or dogma. Such dogma is often complex and detailed, which makes it hard to retain and transmit. Moreover, the lack of emotionality makes them less effective for group cohesion and more prone to being forgotten because they don't motivate a community to practice them together through the same emotion. Doctrine is open to simplification and change through people forgetting the complex detail. The result is that doctrinal rituals must be performed often for the representations to be retained. Unlike rituals with high sensory pageantry, these rituals enable transmission through the consistent and regular presentation of the ritual.

A challenge outlined above is that distinguishing religious ritual from the other rituals is problematic, but another challenge is the reliance on self-contained mental mechanisms. A central notion that I return to throughout this book is the idea that CSR often rests its claims on the workings of self-contained mental mechanisms. The notion of religious ritual is no different because, as noted by McCauley and Lawson (2002; see also McCauley, 2011), religious ritual works as a transmission tool because it fits with cognitive mechanisms that are already in place. They argue that transmission of religious beliefs relies upon the generation of representation, retention of representation, and communication of representation. Transmission is taken to be impossible without a universal cognitive architecture that supports representations. The literature consistently presupposes such mechanisms. Cognition may be emergent in minds (plural) and not in any individual mind in isolation. A transition that I suggest for CSR is to cease (1) talking about cognition employing symbols represented solely inside individual heads to (2) cognition emergent in language (symbols) as a much more sociolinguistic phenomenon integrated with the world. Does CSR have anything to gain from paying attention to cultural psychology and enriching its treatment of ritual?

Why CSR should pay attention to cultural psychology

Thagard (2005) is a generally accepted authority in cognitive science, and he outlines criteria for evaluating the efficacy of an approach to cognition. One criterion is what he calls "Representational Power." This criterion asks us to consider if a given approach to cognition bypasses important functions or information that minds process. My position is that CSR stands to gain representational power if it takes the language practices seriously. Consider how Spelke and Kinzler write about the capacity for belief in terms a "number of separable systems of core knowledge. New, flexible skills and belief systems build on these core foundations" (2007, p. 89; see also Sperber & Hirschfeld, 2004). In this view, religious belief is facilitated because of mechanisms that enable one to hold information about super-agents as they are conceived in one's milieu (see also Barrett, 2007).

As I have outlined above, such a view tends to bypass central components to religious cognition such as the content of belief and embodied experience entailed therein. What the sentences used by Keleman and Rosset (2009) meant and how such beliefs are experienced with compelling verisimilar objectivity has little relevance in CSR because the mechanisms underlying the possibility for belief are treated as primary. How participants came to such beliefs is presumed because CSR would treat them as emergent from cognitive architecture where religious rituals deposit information. When I argue that, to function and succeed in life and be religious in a way like many research participants were, one must be fluent in the language of one's community: the small rituals of life and not just those demarcated as religious. I am putting forth the claim that *belief* means to be fluent in the language of one's community as far as one performs a wide range of rituals. The participants' promiscuous teleology when pressured to perform is a kind of ritual use of language that is missed if it is bypassed in favor of the operation of core knowledge because the very things that belief entails are missed. Promiscuous teleology, for example, is taken by Keleman and Rosset to be about a general non-specific super-agent. When we consider what matters to people who are religious is the development of such belief into a particular content, Christian parents would tacitly socialize a child's promiscuous teleology to a specific belief in Jesus. Such participants would respond in this manner if there was time to think carefully just like the participants in the study did not show promiscuous teleology when they were not under strain. Keleman & Rosset set up two kinds of rituals requiring the enactment of different cognitive performances. That is, the content is central for understanding religion because it is part of the reality experienced and thought about by participants, and the kind of rituals we instantiate are also central. Rectifying the bypassing of such complexity by broadening the approach to ritual could seriously enhance representational power.

Thagard (2005) establishes another criterion for evaluating theory as "computational power," which refers to how well an approach deals with high-level thinking and activities like problem solving (planning, decision making, explaining) and learning. It is about the ability of theory to grasp the complexity of human cognition that we see in the workings of higher-order executive functions. Because of the complexity of human sociocultural contexts and the complexity of our everyday lives, humans need to solve complex problems. Participants must, at times, account for why the world works in the way it does and learn to anticipate what will happen in the future in order to successfully get along in life. An approach to cognition should account for how humans learn insofar as humans dynamically adapt to life in very sophisticated ways. Consider how participants in different experimental conditions adapt to the differential demands of different rituals. To articulate how CSR has room to enhance the computational power, I will elaborate two common examples in CSR.

One example is a system that enables theory of mind (e.g., Barrett, 2012; Barrett & Richert, 2003; Bering, 2006, 2011). It is speculated that children develop socially by being able to solve problems about anticipating others' thoughts and acting on the basis of a mechanism that enables them to imagine the perspective of another person. Researchers in the CSR point out that we never actually have access to the subjectivity of someone else and so we are stuck with something like an imaginative representation of another mind. This capacity can be generalized beyond humans that we concretely deal with from day to day. The logic is that the ability to imagine what is in the mind of one's mother can be generalized to presuppose what would be in the mind of one's deceased ancestor. On the basis of this mechanism, humans are supposedly enabled to generate representations of minds that may not be embodied.

I argued that it is through co-regulation into generic linguistic and cultural rituals that an infant learns the experience and concomitant conceptual organization of ontological givens. Religious cognition is about problem solving in the context of participation in a set of affective practices about how the world is experienced. Such practices are situated in a milieu of normative expectations that one must competently perform in order to be considered authentic. It is not only about a subjective representation based upon maturing cognitive capacity because it is the formation of cognition in concert with others through recursive everyday rituals. In dynamic rituals with others, children form a way of seeing the world before they express themselves in what we normally would think of as language. Formation of cognition happens in infancy in the embodied give and take of life as caregivers speak with a child. A child is brought into participating in socially constituted cognition that shapes the world through these interactions. Instead of treating Keleman and Rosset's (2009) participants' cognition as something that we need to guess to figure out, a potentially more computationally powerful approach is offered by way of showing how we participate in cognition with them and so find the experimental design intuitively sensible (for discussion see Robbins & Aydede, 2009). Familiarity with the ritual use of language entails the familiarity with the psychological performance of religious belief: ritual and language put people in the same mind. Other minds are not a problem on a culturally informed account, and I offer a potentially more parsimonious account of religious cognition as it plays out in higher-order cognition.

Another example is a supposed system enabling the capacity to detect agency (e.g., Barrett et al., 2001; Boyer, 2001; McCauley, 2011). Drawing on evolutionary psychology, researchers have argued that, of the vast array of stimuli in an environment, things that humans need to pay attention to are signs of agency. If one is going to engage in a high-level cognitive process of deciding what to notice, it is advantageous to notice the movement of a potential agent because it may be predatory in nature. CSR has adopted the position that humans are endowed with an agency detection device and that it is overly sensitive as, from an evolutionary perspective, false positives are the safest route to take. As such, humans are considered predisposed to signs of agency in the world around them. Along with the supposed capacity to imagine disembodied minds, a predisposition toward agency can exacerbate this hypersensitive tendency and so set human cognition up to be optimally prepared for belief in a super-agent.

Taking the ritual use of language as constitutive of cognition means that mature functioning is to experience and conceptualize the world in a generic communal manner. It is participation in the way a community shapes the world when members talk about it and is cultivated in children from infancy. As such, the development of religious belief is about coming to apprehend categories of experience in a generically similar way as others. What matters in the study of belief is that children learn to enact the particular world lived in a community. Such claims offer generative potential in reference to religious cognition. Understanding higher-level thinking, like problem solving, requires that the culturally and linguistically shaped categories of experience are addressed because such experience is what determines stimuli (see Thompson, 2007). That is, that which counts as stimuli that could be considered agentic (or not) is shaped by what we have learned through the everyday rituals of language. When selecting features of the environment to think about, we need to realize that the environment is already shaped by language. Computational power can be seriously enhanced if the constitutive role of language in Keleman and Rosset's (2009) sentences is meaningfully taken into account because it offers related insight into what the hyper-active agency detection device is predicated upon. This approach would give us insight into where such belief would come from, and so we can see it through participants' eyes, which is an impossibility from a self-contained approach focusing on psychological mechanisms.

Another criterion for evaluating approaches to cognition is identified by Thagard (2005) as "practical applicability." This criterion is about how well an approach to cognition can be used to improve the life of people. A discussion of representational and computational power above already illuminates how CSR has room to be more practically applicable. Neglecting the everyday rituals of language use cuts off the discipline from effectively addressing religious experience and the content of belief – leaving a science effectively mute to religiosity. Understanding the world, as it is constituted by people in language, both enhances computational power and turns researchers to the practical doings of life. Thagard notes practical applicability also illuminates how people learn, and an example will illuminate how this CSR could stand to be improved on the basis of this account.

Consider how religious cognition is taken to be transmitted over generations through religious ritual (narrowly defined). Researchers in CSR claim that humans have attractor mechanisms that prepare them for some kinds of representations over others (e.g., Atran & Norenzayan, 2004; Barrett, 2008; Gregory & Barrett,

2009; Barrett et al., 2009; McCauley & Lawson, 2002; Pyysiäinen, 2002). Scholars describe how super-agents tend to have counterintuitive elements to them. For example, an ancestor spirit has a mind and this is normal to people, but they do not have a body, which is counterintuitive. Super-agents have a mix of properties where some are counterintuitive and some are not. What researchers in CSR have claimed is that we have cognitive architecture that puts limits on the degree to which a super-agent can be counterintuitive. If a religious ritual involves a super-agent that has no counterintuitive properties, it is not remembered as a viable super-agent and if it has too many counterintuitive properties, then it also dropped from memory. The idea is that people are endowed with mechanisms that are primed to store information about super-agents when it fits within these parameters. Religious rituals that capitalize on these parameters enable transmission of religious beliefs. As such, humans are supposedly endowed with a capacity to imagine others' minds and an overactive agency detection device that capitalizes on this capacity produces a representation of super-agents. Added to this is a mechanism that specifies what sorts of religious representations will be retained and transmitted.

Turning again to the everyday cultural and linguistic rituals can illuminate how there is space for growth in this domain. I argue that it is through everyday rituals that an infant learns the experience and concomitant conceptual organization of ontological religious givens: the conceptions that show up in Keleman and Rosset's (2009) work on promiscuous teleology. A cultural approach specifically focuses on the micro-dynamics of human interaction that specify learning. It provides us with an articulation of how learning of religious cognition occurs in its emphasis of the embodied give and take of life as caregivers speak with a child. A child is brought into participating in socially constituted cognition that shapes their world through these interactions. The potential for explaining how we learn religious cognition does not rely on supposed attractor positions or specifying a counterintuitive detection mechanism. It gets us away from the problem of trying to define what counts as counterintuitive without reference to normativity. A culturally informed approach offers a more parsimonious approach that is more practically viable because it incorporates normativity.

Reframing how CSR approaches such systems in light of the cultural-linguistic constitution of cognition can enhance its representational power, computational power, and practical applicability. All of these criteria are ones that are part of the community of cognitive scientists, which means that I am working on developing an approach palpable (yet provocative) to CSR. It is possible to capitalize on these opportunities in CSR if researchers re-tool how they approach their work by transitioning to more powerful underlying theory. Specifically, it would mean to reformulate research in ways that are not reliant on self-contained processing modules. Such action means that it is possible to develop theory that resonates with CSR and so meets James' truth criterion of fitting with a community.

Moving forward with dynamic systems theory[2]

Interpreting development through a lens that includes the foregoing is not unprecedented in cognitive science (e.g., Maturana & Varela, 1980; Thompson, 2007; Robbins & Aydede, 2009; Varela et al., 1992). Dynamic systems theory (DST) seeks to overcome facile binaries and enable a broader theory of cognitive development (Thelen, 2005). It has led researchers to question the linearity of development (Elman, et al., 1998; Lewis, 2000; Thelen, 2005). There are, of course, milestones and plateaus that are normatively stable across developing children, but there are also discrete differences in how individual children arrive at the same abilities. Thelen and Smith (1996) demonstrated that infants develop abilities through diverse trajectories, with each infant having a unique constellation of experiences. They showed how a biologically programmed pattern of maturation located in the brain is not an impetus for the acquisition of various abilities. Rather, patterns of behavior emerged through ongoing and interactive attempts at a task.

Consider the example of walking where continual interactions among various dynamic systems all contribute to the acquisition of effective bipedal locomotion. An infant's immediate systems also come into play, including body weight and proportions, nervous system and metabolic processes, a child's motivation to walk, and previous experiences with walking behaviors. Each of the components intertwines into a multidirectional flow that creates a *unique* developmental path to a common developmental milestone. Significantly, an infant's system will not autonomously negotiate muscle movement and attempts at equilibrium necessary for walking without first being put in a position, usually by an adult, to practice these competencies. Importantly, Thelen and Smith (1996) observed the development of the capacity to walk as interdependently entwined with an array of environmental contingencies in an infant's milieu. This argument illuminates how dispositional qualities of cognition are instilled through dynamic interaction. The body is coupled to the world insofar as somatosensory systems are entwined with the dynamic forces and happenings in the world (Cromby, 2015). As gravity pulls on a child, her nerves and neurons fire as they are coupled with it. This coupling could be conceived as a starting condition in a recursive system. In relation to the world-body-brain coupling, neurons fire in response that stimulate dynamic movement such as musculoskeletal coordination and attempts at balance. These responses are like finishing conditions that may not result in being optimally balanced, which means that the response is not effectively coupled to the initial conditions. A recursive cycle with different world-body-brain configurations of responses occurs. Eventually, children move to a somewhat effective coupling with a digit-first gait, and this involves their system organizing itself effectively to the dynamic demands of the world. Improved coupling with the environment and increased musculoskeletal sophistication emerges into the heel-strike gait that demarcates skillful human walking.

DST also provides an approach to the emergence of cognition as a phenomenon emergent in dynamic interactions among internal systems of an organism such as everyday ritual use of language enacted by caregivers and physics like gravity. As systems entwine, an infant begins to self-organize based upon the constraints imposed by others (Thompson, 2007; van Orden, et al., 2003). For example, a young child's desire for something would be an instance of self-organization on the basis of neural functioning, caregivers' guidance, objects, and events in the environment. Many other factors can contribute to cognition, altering the course of its organization and function over time. These can include the history of interactions the infant has experienced that accumulate in healthy neurobiological development, the manner in which a caregiver introduces concepts in language, the cultural-historical significance of objects in the environment, and so forth. There is no hierarchical value in this paradigm as each component of the various systems works in tandem to produce a resulting ability or behavior. This also contributes to the non-linearity of psychological development, as small changes in any of the systems can lead to larger subsequent changes in the self-organizing process. Since this process is dependent on the participation of multiple systems, which can include biological, temporal, and ecological ingredients, teasing these systems apart as distinct causal mechanisms becomes an ineffectual task.

Consider how the foregoing would apply to the development of promiscuous teleology emerging in tandem with the linguistic rituals involved in "no." If a caregiver consistently says "no" when an infant reaches for a particular object, an infant will begin to organize itself in the presence of such patterns around this behavior. Internal systems could include an infant's physical (e.g., somatic and autonomic nervous systems) and emotional feeling (e.g., central nervous systems) in tandem with a caregiver's visual and audible actions in recurrent use of the word "no." External systems encompass factors like the cultural significance of the object being reached for. If it is constituted as dangerous, then "no" may be more forceful or panicked accompanied by arousal (e.g., sympathetic nervous system arousal). Other external factors include a caregiver's own conception of how an infant should be interacted with. A caregiver may gently remove an infant from the vicinity of a prohibited object, whereas another caregiver may use physical intervention to stop an infant from touching it. A caregiver furnishes a dynamic context that shapes what an infant can do and feel in a way that co-constitutes an infant's cognitive dispositional experience, and an infant's system organizes to integrate these experiences presented to it.

When caregivers' responses to an infant's actions become predictable, an infant may infer these responses to other actions and "test" them accordingly (Thelen & Smith, 1996). An infant who is consistently removed from crawling under the dining table may test the generalizability of a dispositional self-organization by crawling under the coffee table. If a caregiver is unresponsive in the latter case, an infant will learn there are some behaviors that will elicit a "no" and others that will not. Thus, it is through predicting, testing, and reformulating rituals that an infant learns to organize her own dispositions and so navigate an environment efficiently and effectively. If a context does not support an action, an infant will reorganize behaviors in a way the context supports, such as crawling under the coffee table instead when it learns under the dining table is restricted by a caregiver.

For a caregiver, such dynamic rituals already include an orientation to the ever-present other by virtue of the structure of how language is used. The nuance of the above includes the presupposition of what looks like promiscuous teleology in the expression of language that is oriented to an other. As an infant progressively self-organizes in conjunction with those around him, it will include this self-organization in line with such an ought and promiscuous teleology. The word "no" can be highly variable in its usage and richness such as tacit teleology is part of it. It is through its embodiment that the concept "no" and its emotionality becomes differentiated in tandem with higher order cognition like thoughts about an unseen agent. While an infant may have unique and variable encounters with "no," a consistent caregiver dialogue serves to shape these experiences into a perceptual concept that resonates with those in a community. In this way, religious cognition is developmentally adaptable. It allows an individual to cognitively self-organize in concert with religious language practices and so offers a unique theory of cognitive development that includes higher-level cognition as well as concomitant dispositional cognition. We can come to a place where the sentences in Keleman and Rosset's (2009) study can be understood without bypassing its culturally constituted content and so satisfy communities of religious believers. Simultaneously, we have a sense of the ontogenetic development of religious cognition that satisfies CSR researchers.

CSR often maintains that the development of religious cognition is biologically pre-determined and occurs on a relatively stable maturation trajectory (e.g., Barrett & Kurzban, 2006). As an individual develops over time, additional aspects of a religious cognition blueprint are taken to become operational, allowing for more complex functions. As I will discuss in the next chapter, these functions are asserted to be domain-specific in that particular mechanisms were biologically programmed over the course of evolution to process and organize specific information (Cosmides & Tooby, 1994; Spelke et al., 1992). While this strong nativist stance acknowledges influences from sociocultural factors, these influences are defined as marginal supplements to an already fixed path of cognitive development – sociocultural factors are taken as a stimulus input to cognitive mechanisms that already have the aptitude to perceive and decipher incoming information.

In contract, DST provides a theory of cognition that fits well with the notion of the everyday ritual use of language outlined above. A central issue that drove the need to consider language is the way in which CSR could improve its representational power, computational power, and practical applicability. To get at the phenomenon of religious cognition in a way that realizes these considerations, I argued that researchers can draw on DST to account for how, to function in life, one must be fluent in the language of their community. I also argued that mature functioning is to experience and conceptualize the world the same generic way as members of a community.

DST presupposes the necessity of caregivers using language to shape cognition and consequently, religious cognition. I established how dynamic co-regulation implied that development of religious cognition as a sociocultural phenomenon is enabled by language. The everyday rituals of language use involve developing in a way that is generic to a community, such as acting and talking in ways that are accepted as legitimate and sensible. Participation thereby means enacting language in a way that shapes the perceptual categories, emotion valence, and ideological compellingness generally demarcating a community. The strong form of this claim is that one has developed to the point of shaping the world of experience in the same way as others. The dynamic self-organization that is specified in DST amounts to participation as spelled out above. Self-organization is finding resonance with one's milieu, and this activity means to enact the same world of experience as others. It means that one embodies the same kind of experience as others and to organize oneself to categorize and feel about the world as one ought. Participating in a community by learning its language to shape ontological givens is the same thing as self-organization.

I also argue that researchers need to account for how it is through co-regulation into generic language practices that an infant learns the experience and concomitant conceptual organization of ontological givens. A cultural approach relies heavily on the dynamics of interpersonal relationships and what we referred to as co-regulation. As an infant acts, caregivers respond and in this give-and-take we get the development of abilities to participate in a community. DST treats everyday ritual use of language as central because it points out that all behavior happens in the context of dynamic relations. Acquiring dispositions and enacting higher mental functions requires interrelations among many parts. By arguing that isomorphic approach to developing behavior is unnatural, DST takes an approach toward promiscuous teleology in religious cognition, which immediately presupposes co-regulation.

Conclusion

It is beyond the scope of this book to thoroughly rework previous research by reinterpreting CSR findings through the lens of DST. I will, instead, conclude with other illustrations in reference to theory of mind (e.g., Bering, 2006), agency detection processes (e.g., Barrett et al., 2001), and the retention of religious ritual (e.g., McCauley & Lawson, 2002; Pyysiäinen, 2002).

I described how one system that is taken to enable religious belief is *theory of mind* (e.g., Barrett, 2012; Barrett & Richert, 2003; Bering, 2006, 2011). On the basis of this mechanism, humans are supposedly enabled to generate representations of minds that may not be embodied. DST starts from a place that does not consider the central problem to be one of reading other minds that one cannot access. Because psychology is a matter of self-organization through interpersonal action, one coordinates jointly with others to establish psychological activities in situation. That is, people come to a situation with a wide repertoire of potential to organize themselves in many different ways and dynamically do so in order to functionally couple with others. What looks like theory of mind is actually finding language that resonates effectively with others. Instead of predicating CSR on research that presumes theory of mind, researchers should look to the dynamic everyday rituals as they are embedded in an interpersonal context. The central agenda would include looking at the micro-dynamics of rituals that illustrate how belief is jointly accomplished. It would involve careful consideration of the dynamic interacting systems ranging from biological systems like the parasympathetic system to grand narratives that all jointly come to play (see, for example, Xygalatas, 2012).

A second system that comes to bear when researchers in CSR study belief is the capacity to *detect agency* (e.g., Barrett et al., 2001; Boyer, 2001; McCauley, 2011). Humans are taken to be overly predisposed to signs of agency in the world around them. DST points out, in contrast, that developing cognition includes developing predispositions to act and then adjusting performance in relation to how well (or not) a disposition fits with the milieu. Human cognition is constantly shaping and reshaping as they dynamically attune to this milieu. As such, an immediate startle response to a supposed sign of agency – like a rustling shrub – is a system initially readjusting to the milieu. Research could involve looking at such moments when startle responses are coupled with ineffectual attunements. It would involve, for example, considering the moments when one's attunement to religious cognition being practiced are not aligned or mismatched. Religious cognition could be understood as the dynamic attunement to others in the context of moments when systematic dispositions do not fit and misalignment would occur. Consequently, research would involve considering language and the grand narratives that play a role in the cultivation of dispositions that can be violated.

A third system that furnishes CSR with an explanation pertains to how beliefs can be *transmitted over generations*. Researchers in CSR have supposed that humans have attractor mechanisms preparing them for some kinds of representations over others (e.g., Atran & Norenzayan, 2004; Gregory & Barrett, 2009; Barrett et al., 2009; McCauley & Lawson, 2002; Pyysiäinen, 2002). DST would approach this phenomenon in terms of the capacity of a person to adjust to a milieu. If an idea is present that falls too far from the parameters that someone is accustomed to, it will require too much for them to adjust. Beliefs are transmitted through generations because successive generations are accustomed to those that are practiced. It is in this context that DST theorists talk about the natural drift of a population: cognitive structure (i.e., self-organization) changes slowly over time via small, systemic changes that eventually amount to large-scale change initially too far afield to fit within a person's scope of possible self-organization (Thompson, 2007). Future research could involve the same kind of analysis as above that carefully watches for breakdowns in dynamic systems and readjustment to fit the changing milieu. This work could be conducted through, again, careful observation of beliefs as they are expressed and embodied in language.

Combining a strong form of cultural psychology that attends to language practices with dynamic systems theory offers substantial potential to enhance CSR so that it can address the meaning of sentences like those we see used by Keleman and Rosset (2009). Taking the constitutive role of language and culture seriously enables an approach to cognition that does not bypass the content of belief or the experienced ontological givens entailed in religiosity. An implication is that co-regulation precludes the necessity to concern oneself with finding an initial blue-print or design. Hard-wired maturing mechanisms are not needed when we give primary place to a cultural account. To look to the cognition entailed in religious cognition is to look to the language used by believers. We can move away from approaches that rely upon some sort of primal cause – genetic or otherwise. We can get to an approach that enables us to see cognition in a cultural light and so get at the phenomena of religious cognition in a way that offers more representational power, computational power, and practical applicability. In short, researchers in CSR can move forward in a way to better find belief, and I extend this agenda in the next chapter by turning the third criterion for truth: generativity. The next chapter extends ideas herein by dealing more closely with evolution and experience.

Notes

1 I acknowledge the contributions of two students with whom I worked regarding several themes in this chapter: Michelle McLean and Christina Ashely.
2 For a more elaborate discussion, see McLean et al. (2016)

Chapter 4

Don't throw the baby out with the bathwater II

Evolution and experience

Abstract

Chapter 4 is the last chapter presenting James' approach to scientific truth and I focus on how a claim is true if it is generative: allowing people to move onto new ideas as opposed to friction and tension. CSR takes the position that culturally specific expressions of religious beliefs exist and persist when the beliefs are compatible with evolutionarily formed universal cognitive architectures. This is an example of an a priori paradigm that is problematic because it leaves researchers with little capacity to approach the way that religion involves the constitution of worlds of experience, and this limits the generativity of their work. I outline how CSR predicates itself upon evolutionary psychology and so reduces culture to a phylogenetic feature of the environment. The ontogenetic role of culture is bypassed due to a focus on domain-specific processing mechanisms that are largely irrelevant to religious believers. I propose how we can redirect this focus to generate new directions in CSR. Generative new directions are possible with an enactive approach because it enables paradigm harmonized phylogenetic development with a meaningful ontogenetic role of culture and language. Ultimately, it means that we can research the phenomenologically immediate realities like we see in religious believers and sustain scientifically efficacious theory.

Introduction

A friend was telling me about her three-year-old son named Victor. Victor would demand that streetlights come on or that shadows that displeased him disappear. She found this funny as she was required to share with him that there are things in the world over which we have no agency. I was immediately struck by something more interesting: how *theoretically* impossible his requests were. I was struck by how unlikely it is, according to generally accepted theory CSR, for him to locate such agency in humans. As I touched on in the previous chapter, theorizing in CSR generally tends toward the claim that children are dispositionally prepared to see happenings in the world as caused by some generic unseen super-agent, in contrast to regular agents like animals and people (e.g., Barrett, 2012; Keleman & Rosset, 2009). Victor *should* have attributed the streetlights and shadows to some generic super-agent. What I saw was the opposite: unseen agents like other people – by virtue of their non-presence – who built the lights. Victor located agency in people, and his mother located it in generic absent personae. Where did he get this presupposition of agency, and why was it the opposite of what one would expect?

By discussing these questions, we can contribute to CSR and move forward in new directions. In Chapter 2, I outlined how we need to turn to cultural narratives and stories if we are to develop an understanding of how religious phenomena fit with life. Recall that James (1981/1907) argued that a science with such a fit is insufficient on its own. A true belief, James (1981/1907, 2011/1909) argues, is one that fits the dynamics of life to generate something for us as we move forward such that it makes future activities smoother. A true belief helps us succeed in life as opposed to creating friction and tension. CSR researchers have been moving forward together insofar as their work has generated lots of ideas and discoveries. Unfortunately, there are times where religious believers and researchers have come to a place where things aren't going anywhere so we should "just live with it" by going down our own paths. The problem is that it has not been generative for religious believers due to abstract and disembodied naturalism. The goal in this chapter is to further broaden the way we approach CSR so that we can move forward in a broader range of circles. This chapter outlines potential new directions at a general level that foregrounds phenomenologically lived experience.

CSR is an interdisciplinary approach that attempts to find universal cognitive platforms for religious phenomena (Barrett, 2007) and often leans on evolutionary psychology (e.g., Cosmides & Tooby, 1992, Hirschfeld & Gellman, 1994) to accomplish this task. The general idea is that there are a multitude of universal processing mechanisms that enable religious phenomena when they concurrently operate. This approach to the mind supports the position that universal cognitive mechanisms undergird cross-cultural religious diversity and various beliefs in super-agents (e.g., Barrett, 2007; Boyer, 2001; Cohen et al., 2008; Guthrie, 2002; McCauley, 2011). Culturally specific expressions of religious beliefs are taken to exist and persist when the beliefs are compatible with evolutionarily formed universal cognitive architectures.

There is a significant challenge in presupposing the operation of universal (a-cultural) cognitive architecture underlying religion. Reber argued for an "overabundance of papers focussed solely in exposing and rejecting religious, spiritual, and supernatural truth claims" (2006, p. 195). The challenge is not that the findings are problematic, but a priori, paradigms bypassing religious realities lead to an incomplete psychology. An aspect of life that is central to many people's experiences is simply bypassed, and psychologists need to recognize that the *very realities* religious people embody may be different from that of many researchers.

After all, in contrast to the realities that preoccupy researchers in CSR, many people's ostensive realities include powerfully relevant theistic beliefs and a commitment to the naturalist form of science that bypasses these realities. I am arguing for an approach that addresses the way that humans culturally and linguistically constitute the verisimilar realities in which they exist. CSR has no means of approaching the way that religion involves the constitution of worlds of experience that include the seemingly irreducible reality of super agents. Even if such ideas seem foreign to researchers, such phenomena are precisely what ought to be discussed if research in this field will have any pragmatic value.

This chapter extends Chapter 3 with a continued elaboration of an approach to cognition that can account for the worlds of experience constituted in language and culture; the worlds that laden with the meaning and experience of phenomenollogically immediate realities. The first portion of this chapter attempts to outline how CSR predicates its approach to mind upon evolutionary psychology. I raise the problem of how this appropriation of evolutionary psychology entails the inclusion of culture as a phylogenetic feature of the environment, which means that the ontogenetic role of culture is effectively bypassed. The reason for this bypass is that CSR focusses on domain-specific processing mechanisms, and this focus must be redirected in order to generate new directions in CSR. By drawing on some promising work in CSR, I then outline how culture and cognition are irreducibly interdependent in the ontogenetic development of religious cognition. An approach to cognition called Enactivism is the focus of the latter third of this paper because provides a paradigm of cognition that harmonizes phylogenetic development with a meaningful ontogenetic role of culture and language. My aim is the thereby offer a way to work within the phenomenologically immediate realities like we see in religious believers while maintaining a scientifically efficacious theory of cognition. In so doing, we can grasp why Victor perceived agency and what it has to do with religious belief.

CSR, evolutionary psychology, and domain-specific processing mechanisms

As mentioned above, researchers in CSR have tended to position themselves within a debate over whether or not universal cognitive mechanisms undergird religious diversity (e.g., Barrett, 2007, p. 1; Boyer, 2001, p. 3; Cohen et al., 2008, p. 113; Guthrie, 2002, p. 39; McCauley, 2011, p. 13). Such universals are generally taken to exist, and they are presumed to provide a cognitive architecture that enables religious beliefs. Humans, for example, have a propensity to believe in an afterlife on the basis of an pregiven cognitive ability to representation of the content of others' minds in one's own mind (Bering, 2006). This cognitive ability can be extended to represent what would be in the mind of a deceased ancestor – i.e., creating afterlife belief.

Theory in CSR often defaults to evolutionary psychology to explain where such abilities come from (Tooby & Cosmides, 1992; Cosmides & Tooby, 1994; e.g., Barrett, 2004; Bering, 2011; McCauley, 2011; McCauley & Lawson, 2002;

Sperber, 1994). Survival challenges in a Pleistocene environment, according to evolutionary psychologists, produced adaptive modifications that enhance genetic fitness (Cosmides & Tooby, 1994, p. 86). The *only* causal forces that compel a population toward functional adaptations is taken to be selection (Tooby & Cosmides, 1992, p. 57). Phylogenetically recurrent environmental circumstances structure the design of adaptations (Tooby & Cosmides, 1992, p. 61–62, 69). Evolutionary psychologists purport that there is no good reason to argue that selection favored anything other than hard-wired genetic solutions to survival challenges faced by our stone-age ancestors (Cosmides & Tooby, 1994, p. 87). In other words, the structure of the Pleistocene environment caused corresponding adaptive organization of cognition.

In an innovative move, evolutionary psychologists point out that a dualistic nature-nurture view "expresses only a premodern version of biology, whose intellectual warrant has vanished" (Tooby & Cosmides, 1992, p. 21). Evolutionary psychologists therein attempt to distance themselves from being open to the charge of being genetic *or* environmental determinists. The environment is taken to influence the shape of human cognition through natural selection and the evolutionary emergence of a phenotype relies upon interaction between genes and environment. Evolution "shapes the relationship between genes and the environment such that they both participate in a coordinated way in the construction and calibration of adaptations" (Cosmides & Tooby, 1994, p. 86). They avoid being called biological reductionists by taking the position that the environment – including culture – impacts human cognitive architecture over phylogenetic time.

Note how the following quote points to a problem with the theory:

> culture is a manufactured product of evolved psychological mechanisms situated in individuals living in groups. Culture and human social behaviour is complex and variable, but not because the human mind is a social product, a blank slate, or an externally programmed general purpose computer. Instead, human culture and social behaviour is richly variable because it is generated by an incredibly intricate, contingent set of functional programs that use and process information from the world, including information that is provided both intentionally and unintentionally by other human beings.
>
> (Tooby & Cosmides, 1992, p. 24)

A cognitive system, as the quote infers, provides mechanisms designed to create an organism's ability to respond to the environment (Tooby & Cosmides, 1992, pp. 28 & 116). Without a large number of specialized mechanisms, it is supposedly impossible for a human to respond to the environment. Lying beneath variable behavior like Victor's comment are universal mechanisms and "superindividual interactions depend intimately on the representations and other regulatory elements in the head of every individual involved and, therefore, on the systems of computation inside each head" (Tooby & Cosmides, 1992, p. 47). While resisting the charge of being biological or environmental reductionists, evolutionary psychology grants nearly exclusive primacy to phylogenetically shaped cognitive processes (see Derksen, 2007, 2010). Culture merely works as part of the natural environment to establish mechanisms over phylogenetic time.

On the one hand, researchers in CSR rest their work upon claims that religious phenomena emerge due to evolutionarily designed cognitive architecture. On the other hand, CSR is deeply indebted to domain-specificity (Barrett & Kurzban, 2006; Hirschfeld & Gellman, 1994), which pertains to the limited range of a given mechanism's ability and capacity such that it can only handle narrow amounts of specified information (Hirschfeld & Gellman, 1994). Barrett & Kurzban outline how such domains are "individuated by the formal properties of representations because, we believe, this is the only possible means by which brains systems could select input" (2006, p. 630). This claim means that a given mechanism responds to a particular input that fits its scope. Such a position echoes others who describe domain specificity as "a genetically specified computational device in the mind/brain (henceforth: the mind) that works pretty much on its own inputs pertaining to some specific cognitive domain and provided by other parts of the nervous system (e.g. sensory receptors or other modules)" (Sperber, 1994, p. 40). Organism-relevant input is defined in terms of being relevant to one or more domain-specific mechanisms. When evolutionary psychologists claim that natural selection is purported as the only explanation for functional organization of our cognitive mechanisms, they mean that natural selection creates domain-specific mechanisms (see Cosmides & Tooby, 1994, p. 88). The results are the aforementioned evolutionarily designed constraints placed on an organism's capacity to respond to the environment: ". . . an organism's behaviour cannot fall within the bounds of the constraints imposed by the evolutionary process unless it is guided by cognitive programs that can solve certain information processing problems that are very specific" (Tooby & Cosmides, 1992, p. 97). Whatever we see in Victor is cultural insofar as the Pleistocene environment is taken to have shaped his cognition so that it can process information – the stimuli in the environment are not taken to be constitutive of his cognitive architecture.

Domain specificity is not dismissed on the basis of the fact that people behave in novel ways in new situations. Integration and interactivity of information from multiple sources is expected and, once some mechanisms are formed through natural selection, their output could be input for other mechanisms (Sperber, 1994). Novelty that we see could be explained on the basis of "actual" versus "proper" domains (Barrett & Kurzban, 2006, p. 635). A *proper domain* refers to a domain that mechanism is designed for (Sperber, 1994, p. 50 and following). The selection of input formed in proper domains is formed by representations pertaining to the Pleistocene era. As such, the proper domain refers to formal rules specifying content of Pleistocene era. It is in this manner that Tooby & Cosmides claim that human architectures are " 'pre-equipped' (that is, reliably develop) specialized mechanisms that 'know' many things about humans, social relations, emotions and facial expressions, the meaning of situations to others, the underlying organization of contingent social actions such as threats and exchanges, language, motivation, and so on" (1992, p. 89).

Obviously, humans like Victor do not exist in a Pleistocene environment any more, but they still function reasonably well. *Actual domains* refer to any range of stimuli that can satisfy the proper domain's input conditions. For example, consider Bering's (2010) discussion of the input of a hyperactive agency detection device. This device serves as an example of a domain-specific mechanism often invoked in CSR. It involves a supposed device that is specialized for detecting agent-directed activity as its proper domain. Other input criteria other than just seeing a human or animal serve to trigger this device. A streetlight may turn on and the device supposedly kicks into gear and creates Victor's suspicion of agency. The actual domain is the light turning on. The result is a supposed mechanism that over-produces Victor's perception of agency underlying change in the world, and such ubiquitous agency is theoretically supposed to establish a belief in a super-agent.

Core rules for processing represented stimuli forms the basis for the notion of a "tool kit" and this notion if consistently invoked in CSP. For example, Barret claims that "human mind/brains exhibit a number of functional regularities regarding how to process information. These functional regularities are also known as domain-specific inference systems or 'mental tools'" (2007, p. 2). Boyer similarly asserts that the mind "comprises lots of explanatory devices, more properly called *inference systems*, each of which is adapted to particular kinds of events and automatically suggests explanations for these events" (2001, p. 17). Such evolved mechanisms are taken to enable cognitive architecture that amounts to religion.

Just like evolutionary psychologists attempt to avoid the charge of biological reductionism by attempting to circumvent a naive nature-versus-nurture dualism, researchers CSR make efforts to do the same (see Cohen et al., 2008). The challenge, however, is that this attempt is not wholly convincing in light of the often-repeated phrase that "our modern skulls house a stone age mind" (Tooby & Cosmides, 1992; Cosmides & Tooby, 1997). While evolutionary psychology involves a non-reductionist agenda that makes room for culture in phylogenetic terms, its proponents claim that cognitive architecture is exclusively shaped in the Pleistocene era and is passed on almost exclusively through genetics, which implies something that looks like biological reductionism (e.g., Cosmides & Tooby, 1994, pp. 106–108; Tooby & Cosmides, 1992, pp. 82–87). The trend that emerges in CSR is this conception of culture that essentially by-passes the immediate ontogenetic role that language and culture play. Barrett et al. (2009), for example, argue that a super-agent would be retained in Victor's memory if it violated a few natural expectations but not too many. The cultural content or meaning of a specific agent is irrelevant because it is the number of violations of natural expectations that matter. A super-agent that can walk through walls but has a natural human-like mind is represented in memory as opposed to an agent that violates many assumptions. This is an example of cognitive constraints that come from natural selection. There is a minimal effective ontogenetic role of immediate culture on cognition. A result is a view inherited from evolutionary psychology: the environment that matters to Victor's architecture is the Pleistocene environment and not the immediate situation other than serving to provide input in actual domains.

The bottom line is that culture is rarely given more than a phylogenetic role when cognition is merely conceived as evolutionarily designed domain-specific mechanisms. Where I am calling for an approach that can accommodate the phenomenolgically verisimilar realities that are shaped in language and culture, CSR's tendency to rest upon evolutionary psychology runs the risk of narrowing itself to a paradigm that cannot accommodate such phenomana. Gantt et al. (2012) have argued that evolutionary psychology expands its metaphors to a metaphysical level removed from actual evidence (see also Richardson, 2007). While I am not comfortable speculating about metaphysics, Gantt et al. certainly make a good case that evolutionary psychology perpetuates a particular world view. They show how evolutionary psychologists conceive of reality in a singular sense that reflects a naturalistic paradigm incongruent with human sciences (see also Slife & Reber, 2009). This paradigm excludes any sort of world view that is non-scientific while it is, in fact, a narrow conception of science shaped in metaphor just like any other view of reality. Gantt et al. show how presuming an a priori phylogenetic role for culture and locking cognition inside the skulls of humans is expressive of a paradigm that shapes what counts as evidence and does not represent ideas emergent from observation. In what follows, I will seek to propose a paradigm that can address verisimilar experiences of realities and then offer a theory of cognition that can further generatively expand CSR.

Cultural psychology: humans constitute realities

Previous chapters have addressed cultural psychology to provoke a discussion about how cognition and culture are ontogenetically interdependent (e.g., Bruner, 1990, Ratner, 2006; Shweder, 1991; for overview see Valsiner, 2012). Along with other cultural psychologists, I argue that defining the immediate environment as constitutionally independent of cognition – aside from serving as general actual domain input criteria – reflects a problematic conception of the relationship between culture and cognition. Understanding the realities experienced by religious believers realities requires understanding how cognition is culturally and linguistically constituted because "personal life takes shape in cultural terms, or better yet, perhaps, that individuals are necessarily and continually involved in the interpretive apprehension (and transformation) or received symbolic models" (Rosaldo, 1984, p. 140).

Consider, for example, social constructionism and discursive psychology. Kenneth Gergen (1994) spelled out the role that language and sociality played in the social construction of cognition. For example, he spelled out how what are usually taken to be processing mechanisms pertaining to self-representations are actually a social phenomenon ontogenetically shaped in social experience and language use (Gergen, 1999a, 1999b). In the United Kingdom, discursive psychologists such as Derek Edwards (1997) and Jonathan Potter (Potter & Wetherell, 1987) pursued a similar agenda. For example, Edwards (1997) carefully and systematically outlined how mechanisms enabling categorization are not really explainable in terms of information processing. He showed how perception and categorization are better conceived as social constructions occurring in interactions with others. In a similar vein, Rom Harré (2002) and Jerome Bruner (1990) commented on the cognitive revolution to point out that the idea of processing mechanisms is misguided. They argue that language shapes the perceptual categories and cognitive performance of humans. There are other substantial bodies of work that carefully document the sociolinguistic constitution of what are normally taken to be cognitive properties that cannot be explored in depth here (see Wertsch, 1985, 1991). Suffice it to say that there is much literature demonstrating the viability of an alternative mode of cognition, including alternative textbooks (e.g., Harré, 2002).

This work is expanded by others interested in the *embodiment* of culture and language. A central notion is that, when humans engage language, they align themselves with the folk meanings *experienced* in cultural symbols and practice (Baerveldt & Voestermans, 2005; Cresswell, 2011; Cresswell & Smith, 2012; Cromby, 2015). Children learn language, and to learn a language is to learn the experiential realities that come with it (Cresswell & Teucher, 2011; McLean et al., 2016). To learn to be "afraid of the dark" is to learn that "afraid" is experientially part of "the dark." Work with infants and young children has shown how humans are socialized to bodily experience realities in the same way as caregivers, and this work ties such socialization to language and symbolic expression (Cole, 2003; Fogel, 1993; Rogoff, 2005). For example, Oppenheim et al. (1997) examined dialogic interplay between mothers and their children. They showed how dialogic interplay between mothers and children constituted children's ability to regulate their own behavior and emotion. This line of work expands social constructionism and discursive psychology insofar as it shows how, when humans engage in sociolinguistic practices, they align themselves with the folk meanings *experienced* in sociolinguistic symbols.

The forgoing, despite some differences among theorists, directs our attention to the way that acquiring an embodied mode of being that gives phenomenolgically experienced realities compelling force is enculturation and language acquisition. Participating in a religious community, for example, involves language and culture that shapes the world as it is experienced with the verisimilar reality – the realities within which religious participants live and what I seek to address. The point is that culturally constituted experiential meanings are centrally important and not domain-specific processes. In contrast to the idea of domain-specific processing mechanisms where the authors "do not intend a reading of domain as *content domain*, in the folk sense of domains individuated by the *meaning* of their constituents" (Barrett & Kurzban, 2006, p. 630, emphasis added), folk meaning is central in a culturally oriented view. The claim that "personal life takes place in cultural terms" is meant in a deeply pervasive sense where culture and cognition are constitutionally interdependent in ontogenesis.

Consider how CSR misses religious meaning and experience in Victor's employment of agency because it misses the co-constitution of culture and cognition. An extreme example of this problem can be seen in authors that purport a worldview that defines immediate religious contingencies as epiphenomenal. For example, Guthrie described religion as a "bi-product of . . . a perceptional and cognitive strategy" (2002, p. 39). He meant that the culturally immediate specifics of a religion are not relevant because cognitive mechanisms are of interest. Other authors make similar claims with an emphasis on "[c]ore systems for representing objects, actions, numbers, places, and social partners . . ." (Spelke & Kinzler, 2007, p, 90). Boyer (2001) regularly refers to "inference systems" to address evolutionarily shaped domain specific processing systems. A similar approach is found in Bering (2011), who wrote that "we are indeed contained entirely within our own skulls" (p. 23) and later that "the human brain, like any physical organ, is a product of evolution, and since natural selection works without recourse to intelligent forethought, this mental apparatus of ours evolved to think about God quite without need of the latter's consultation" (p. 43). Bering was interested in showing how our minds have pre-given knowledge independent of the immediate environment. He draws on authors such as Keleman who argue for a "promiscuous teleology" and this concept pertains to a general propensity to explain the realities as having a purpose ordained by an agent (see Keleman & Rosset, 2009). In contrast to Victor's behavior, CSR would have predicted that the shadows and lights were driven by a super-agent. Bering seeks to describe a "universal ontology of cognitive abilities" (Bering et al., 2005) and expresses a general trend in CSR to promote cognitive architecture formed by domain-specific processing mechanisms. The notion of actual and proper domains, moreover, effectively creates this split and implies a schism between cognitive faculties and the environment that simply does not fit what cultural psychology has shown to be human. The folk meanings and experiences that are the fabric of language and culture are bypassed because of the way culture and cognition are separated. Isolating cognition from experienced realities betrays the way that human cognition is entwined with a milieu (see Peters, 2013; Tallis, 2011). In a culturally informed approach, a stimulus is constituted by sociolinguistic practices that constituted experiential reality. When cognition is not treated as principally distinct from the world, stimuli are not distinct from the organism. The meaning of a stimulus is shaped by the individual as she has participated in life with others.

We need an approach that can handle phenomenologically immediate experience in order to grasp religious beliefs and the realities that they entail. Rather than dismissing folk psychologies as irrelevant, I propose a science that is culturally informed and takes seriously the constitutional force of language can be more productive. For example, as I discussed in Chapter 2, Belzin (2010a) approached a Christian tradition with an attempt to apprehend what this meant by using the language of salvation and sanctification. He addressed such terms through participating in the lives lived by those in this particular tradition to understand how they are experienced. In so doing, he studied the sociolinguistic constitution of this community's cognition. If we follow Belzin, "we must appreciate the ways that such understandings grow, not from an 'inner' essence relatively independent of the social realities, but from experience in a world of meanings, images, and social bonds, in which all persons are inevitably involved" (Rosaldo, 1984, p. 139).

Some of the most exciting work in CSR falls in line with the proposal that it be enhanced in a way that can account for the meaning of such culturally constituted realities. Consider how some authors in CSR write that "just as culture does not hover above cognition, so cognition is not somehow insulated from culture. . . . The shaping/constraining effects of evolved psychological mechanisms are only *some* of the many factors influencing religious innovation and transmission" (Cohen et al., 2008, p. 113–114, emphasis added). Moreover, others write that "[u]nlike the social lives of non-human primates, human social life is thoroughly cultural . . . there need not have been anything in the ancestral module that precluded the possibility of culturally enhanced or *constructed* cues" (Sperber & Hirschfeld, 2004, p. 44, emphasis added). Whitehouse (2004), furthermore, goes so far as to write that "human processing is not like a computer" (p. 22). Such quotes indicate that CSR is not necessarily committed to a culturally isolated conception of cognition.

Sperber (1996) is another such instance of evocative work that is potentially generative in the direction that I propose and this is because he attempted to offer an approach to culture that is free of naïve reductionism. Consider how he wrote that "what I want to suggest . . . is that psychology is necessary but not sufficient for characterization and explanation of cultural phenomena." (p. 60). He thereby makes room for context and culture with claims such as "metarepresentations are brain states described in functional terms, and it is the *material interaction between brains, organisms, and environments* that will explain the distribution of a representation" (p. 26, emphasis added). Sperber opened the door to an approach that takes culture and language seriously by initiating the potential to see them as integral to cognition. This potential has not been realized because he still relied upon and is appropriated in a way that includes domain-specific processing mechanisms. For example, he is clear that he seeks to explain cultural macrophenomena as cumulative effects of individual mechanisms and inter-individual mechanisms: "there are only mental representations, which are born, live and die within individual skulls, and public representations, which are plain material phenomena" (p. 81). Where Sperber offered a non-reductive materialism that failed to materialize, I am attempting to redress this shortcoming.

Robert McCauley also raises hopes with similar potential in his more recent work (see McCauley, 2011). Like Sperber, McCauley opened the door to a potential approach that includes an ontogenetic role for culture. He begins his book by arguing that cognition is embedded and embodied (p. 17). Unlike Sperber, McCauley is more forthright about the interdependence between culture and cognition where he made claims such that cognitive accomplishments are not confined to what happens in people's heads (e.g., p. 18). He also claims that cognition does not depend on exclusive or genetically determined capacity and thereby opens the door to a contextual approach that includes a meaningful inclusion of culture. In a mode strikingly similar to Sperber, however, McCauley still reverts to the primacy of domain-specific processing systems where operations are mandatory. As his book progresses, he moves further toward the familiar line of domain-specific processing mechanisms shaped in evolution. He eventually asserts cognition does not depend on anything culturally distinctive and cautions that it is possible to overplay how much something is culturally embedded (p. 37). The same primacy of self-contained processing in Sperber is echoed in McCauley.

I am attempting to highlight how researchers in CSR are sophisticated scholars in their own right and do not make shallow commitments to a perspective of cognition that is independent from culture. Authors like McCauley and Sperber hint at a potentially richer form of CSR that can account to phenomenologically experienced realities like those we see in religious belief. I seek to push the work a little further in this direction in order to explain how it is possible to have a paradigm about cognition that foregrounds the ontogenetic role of culture. An issue to deal with, however, is the role of evolutionary psychology because domain-specific processing mechanisms are supposedly rooted in evolution. By presenting a paradigm of cognition that both sidesteps the need for domain-specific processing and offers an alternative view of the evolution of human cognition, I will propose an approach to cognition that can account for the religious realities that people embody.

Generative cognitive science in CSR: enactivism

Embodied cognition is a unique approach that can broaden the way researchers in CSR approach cognition and evolution (see Karasavvidis, 2002; Noë, 2009; Robbins & Aydede, 2009; Shapiro, 2011). CSR is nested within a paradigm that presupposes cognition to be about happenings rooted in an individual brain. Embodied cognition treats it as belonging to the whole sensory-motor activity of the body. Embodied cognition extends beyond the brain into the whole body "such as the sensory organs, the musculoskeletal system, and relevant parts of the peripheral nervous system" (Robbins & Aydede, 2009, p. 4; Varela et al., 1992), which was a central point in Chapter 3. Cognition can be understood in terms of contextually bound embodied action instead of self-contained processing mechanisms (Noë, 2009). Embodied cognition potentially moves us forward in a way that includes culture and language, but it only goes so far. Embodied cognition involves the body, but it is not specifically focused on phenomenologically immediate experience like we see in religious belief. Like any intellectual tradition, it breeds scholars that attempt to rectify such challenges and enactivism, as promoted by Evan Thompson (2007), does attempt to get at such experience. Thompson is a good source to turn to because his work addresses phenomenologically experienced realities in direct relationship to cognitive science (Thompson, 2007; Varela et al., 1992).

Enactivism and cultural psychology

Enactivism, in faithfulness to the embodied cognition tradition, outlines how cognition is fundamentally and constitutionally engaged in the world outside of an organism's individual body. It treats cognition as dynamic systems achieving sensory-motor organization interdependently with the surrounding milieu. Rather than treating cognition as a feature of the domain-specific mechanisms anchored in the brain that receive stimuli and respond on the basis of such processing, cognition can be seen as emergent in circular causality of continuous sensory-motor interactions involving the brain, body, and milieu (Thompson, 2007, p. 11). At the neurological level, for instance,

> individual neurons do not detect objectively defined features. Rather, assemblies of neurons make sense of stimulation by constructing meaning, and this meaning arises as a function of how the brain's endogenous and nonlinear activity compensates for sensory perturbations.
>
> (Thompson, 2007, p. 530)

Consider, as an example, the idea of how children supposedly represent the contents of others' minds and how Victor imagines the intentions of the man that controls the streetlights. Usually, CSR draws upon this idea that we understand others' minds by forming a theory about what is present within them: Theory of Mind (e.g., Barrett, 2012; Keleman & Rosset, 2009). In contrast, an enactive approach promotes an increasingly attuned organism aligning with other organisms' recursive ritualistic coregulation of everyday life. The sympathetic nervous system of a child may be escalated in an organism like a caregiver, and what this arousal amounts to depends on the context. It could be a feeling of excitement or indignation if there is a cultural and linguistic practice that gives shape to this sympathetic nervous system arousal in a milieu. A child's own sympathetic nervous system may mirror that of the caregiver such that there is attuned arousal (see Greenspan & Shanker, 2006; Chapter 3). Learning is to ontogenetically couple this sensory-motor attunement with symbols and others' higher-order cognition with others. Where it may look like there is a representation of others' mental states, what is actually happening is a child falling into synchronicity with the realities and the caregiver who is also part of such realities. These are the realities into which Victor has been enculturated.

The environment in which a child, like Victor, is situated takes on a different role in enactivist theory. Consider how Thompson (2007) points out that the environment, as such, should not be separated from human cognition in terms of stimuli principally distinct from domain-specific processing mechanisms. The environment is constituted by cognition and vice versa and thereby wrote that "'Inner' and 'outer' are not pre-existing separate spheres, but mutually specifying domains *enacted or brought forth* by the structural coupling of the system and its environment" (Thompson, 2007, p. 26; emphasis added). It is an approach highly resonant with the strong role that I have given to language and culture, and it means that stimuli would not be considered triggering conditions. Dismantling of the dichotomy between "inner" and "outer" blurs the boundary between world and cognition such that the world, insofar as humans can possibly know it, is shaped by cognition. For humans, experience goes beyond affect insofar as it entails value and meaning such that an *environment* with stimuli is treated as communally constituted *realities*. These realities present themselves to us in ways that compel us because they are integral to us and this is how enactivism offers an account of phenomenologically immediate experience that we see in religious beliefs. The world seemingly presented itself to Victor in its culturally constituted experiential richness because experience does not have an object standing over against him as a distinct subject. Rather than claiming that Victor has representational content *about* a supernatural or non-supernatural being, in his *manifest experience*, he engages agents of all sorts as part of reality.

Enactivism accounts for how realities, insofar as we can possibly experience in the full embodied sense, are bound to language and culture. Participation in a community enables people to be tacitly tutored in a reality that those in a community take for granted (see also Tallis, 2011). Consider again the approach to what looks like representing others' minds. An enactive paradigm of cognition involves the attunement to jointly constituted realities. This achievement includes cultivating synchronicity with the sensory-motor dynamics of others and such dynamics as they are coupled to immediate life. A feeling of the presence of an agent, for example, involves a situation where it is appropriate as constituted by a community's sociolinguistic practices. Through a history of engagement with many others, there are normative dispositions as to what situations are appropriate for agents, supernatural or otherwise. As such "(1) perception consists in perceptually guided action and (2) cognitive structures emerge from recurrent sensorimotor patterns that enable actions to be perceptually guided" (Thompson, 2007, p. 173). Realities, insofar as they can possibly be experienced, are constituted by such dynamic mutuality. A generative implication is that Victor's cognition is shaped in its coupling with others in a way that intrinsically includes language and cultural practices that offer him the interpretation of the situation.

Such claims fit with the perspective of cultural psychology set out in the spirit of William James. What Thompson adds is a complementary approach to cognition that explicitly integrates biology (see Baerveldt & Verheggen, 2012). Language is deeply embodied by virtue of how it is coupled to a human organism's sensory-motor engagement with life. There never is a natural reality for Victor over which his experience is laid because it is experienced as laden with culturally and linguistically constituted value and meaning: it presents itself to him in its experiential richness that includes the potential agent that can control the lights. A different constellation of social and cultural systems would be a presupposition that a super-agent could take care of it. Simply put, Thompson further shows how cognition can also be understood as experientially bound to realities constituted by cultural and linguistic practices in which we are embodied participants. This view opens up an approach to the realities within which people, including religious practitioners, live.

Enactivism and evolution

Enactivism puts forward the claim that ontogenetic development is about self-organizing in attunement with others (Thompson, 2007; Varela et al., 1992). The central idea is that Victor organizes his sensory-motor, cultural, and linguistic activities in harmony with others is therein organizing himself. As he moves through different situations with others, a dynamic reorganization occurs. Different constellations of milieus require differently reorganizing one's psychology such that self-organization would include enacting appropriate feelings and expressing appropriate rationality. To organize oneself is to experience the appropriate realities at an appropriate context. As children grow, they become better at organizing their emotions and thoughts and so become better able to participate in the dynamic rhythm of life. It is marked by success defined in terms of acting and perceiving in a manner that is close enough to others and the ostensively given realities that others don't act in a manner to correct one. The kinds of behavior a child enacts are increasingly sensible and appropriate because she is organizing herself in concert with life and enacts it along with others: she sees the realities as others do. A child reaches cognitive maturity when a child is competent in the cultural and linguistic practices and we could call this the ontogenetic development of a functional cognitive system (Thompson, 2007, p. 99).

This approach to cognition provides an alternative account of human evolution that starts with rethinking the notion of adaptation. Where adaptation is traditionally taken to be a capacity that offers an advantage to an organism in passing on its genes, enactivism treats adaptation as an attunement of a human to the realities in which Victor participates with others. An enactive view of adaptation is one that starts from the observation that humans are fundamentally interdependent with others and phenomenologically immediate realities. The consequences are that

> heredity depends not on 'transmission' of genetic information for phenotypic design, but on the reconstruction of patterns in ontogeny, a process that involves many other developmental resources besides genes. Second, the innate/acquired distinction is not applicable to developmental processes. Finally, natural selection is not an external force, but the differential propagation of developmental systems.
>
> (Thompson, 2007, p. 202; see also p. 184)

As realities change with different communities, so does Victor because his adapting amounts to a self-organization in relation to realities lived by others. This position stands in direct contrast to the idea that there is a single human organism faced with adaptive problems conceived as its individual survival. Hence, it better fits the ways in which humans are irreducibly entwined with others. It makes room for how humans are ontogenetically entwined with each other and the socio-linguistic constitution of realities that are experienced with super agents (see Buller, 2005).

The concern that then arises with such a notion of adaptation is about the role of survival pressures. That is, it calls for an account of phylogenetic changes that occur in a population and addressing it requires a discussion of how enactivists approach the notion of reproduction. Self-organization is involved in reproduction in that, according the enactivism, ". . . reproduction consists in one unity organizing another unity of the same class, that is, having the same organization" (Thompson, 2007, p. 167). The dynamic system in a human is reproduced in the sense that another human organizes herself in the same way. We can imagine how, as Victor's mother starts at a rustle in some shrubs along a pathway, he is tacitly socialized into such behavior and experience of the realities. In this example, the dynamic organization that produces what looks like a hyperactive agency-detection device is reproduced in another human as the latter actually self-organizes in attunement with others. Attunement among conspecifics happens in synchrony, and the long-term outcome is chronological attunement. Such a process gives rise to historically linked psychologies.

Population change over phylogenetic time is described in terms of "natural drift" by enactivists (Thompson, 2007, p. 97). Change emerges in the internal dynamics of an organism that is interdependently connected with its milieu and conspecifics in that milieu, which means that change over time via gradual coordinated change in self-organization leads to a gradual change in a population (Proulx, 2004). After all, "[i]n any historical lineage generated by reproduction, there will be both conservation of, and variation in, the structural characteristics of the unities from one generation to another" (Thompson, 2007, p. 169). Instead of genetic traits being selected by environmental contingencies, enactivists point out how organisms that compose a population are able to self-organize to adjust to their realities and this notion of realities integrates basic selective pressures such as food scarcity with more complicated sociolinguistic phenomena. For humans, ideological misalignment with others can be just as deadly as starvation, and enactivism recognizes this important fact.

Change can occur in relation among organisms and their environment as the environment changes in tandem with organisms, but there is also stability over time. Stability is described in terms of robustness and

> *Robustness* is the capacity not to change when conditions change, a capacity for self-maintenance, self-adjustment, and self-organization in the face of change. *Flexibility* is the capacity to change in relation to changing conditions, to accommodate change. Robust and flexible developmental processes, because they allow for phenotypic variations to arise in development that are not lethal to the organism, enhance the capacity of organisms for evolutionarily change.
>
> (Thompson, 2007, p. 195)

Robust systems are established over time, and these are dynamic systems that remain functional as a milieu changes – both in terms of features like weather and terrain and in terms of cultural features. Robust systems don't substantially change when the environment changes because the system is able to reorganize itself to accommodate the milieu. It is such robust systems that amount to a dynamic stabilization and what are misattributed to be genetically endowed cognitive platforms. A trajectory of change is thereby emergent in the interwoven quality of multiple organisms and their environment, but change can be slow on the account that self-organization can be robust. The generativity of this approach is that it effectively brings the ontogenetic efficacy of culture together with phylogenetic evolution.

To return to CSR, an important feature of researchers in area is a commitment to finding universal features of cognition that undergird religious phenomena (e.g., Barrett, 2007; Boyer, 2001; Cohen et al., 2008; McCauley & Lawson, 2002). Traditionally, this search for universals has manifested in attempts to specify universal cognitive architecture composed of supposed domain-specific processing mechanisms. Because this approach sidesteps the phenomenologically experienced realities within which religious believers participate, it sidesteps the actual phenomena of lived religious belief and experience. I have attempted to cultivate an approach that allows CSR to develop a more robust field of work. By expanding upon work already addressed in authors such as Sperber (1996) and McCauley (2011), it is possible to redirect CSR in a way that makes room for an ontogenetic role of culture and language. The realities of experience that are shaped in communities can be approached with an enactive view of cognition. The foregoing has attempted to clear a path with a turn to cultural psychology that addresses agents experienced as irreducibly real. I have also attempted to clear a path by offering a paradigm of cognition that addresses the way that cognition is constituted by culture expressed in sociolinguistic practices that includes a different view of evolution.

What we do with evolution and the implicit issue is one of the search for cognitive universals? First, it is important to be clear how evolutionary psychology and domain-specific processing misses the interdependence of culture and cognition. In the words of Thompson (2007):

> Genocentrism and computationalism. . . [b]oth perpetuate the dualisms of hardware versus software, matter versus information, body versus mind, and both mechanize the role that particular subsystems play in what are fundamentally dynamic phenomena of the whole organism embedded in its environment. . . . A better metaphor for development than 'following coded instructions' is 'laying down a path walking.' This metaphor implies that there is not separation between plan and executed action.
>
> (pp. 174–180)

Any approach to finding such universals requires that this interdependence be acknowledged, but we do not need to fixate on the universals being particular self-contained mechanisms. Authors such as Peters (2013) have pointed out that the notion of genetically determined processing mechanisms is problematic because it contradicts clear evidence to the contrary. He points out that the removal of the ontogenetic view of culture simply misses neurological evidence (see also Richardson, 2007). Other authors, such as Buller (2005), have pointed out more generally that ontogenesis is crucially important in terms of understanding cognition. An enactive account offers a view of cognition that makes room for the ontogenetic role of culture. It takes us out of the realm of treating humans as individuals by recognizing our fiercely interdependent quality. It does so, however, without having to dismiss universals. Thompson is clear that self-organization can lead to historically linked psychologies and so writes about some organization configurations as robust. Religious experiences have had consistent patterns over time and CSR has picked up on such consistencies. McCauley and Lawson (2002) have, for example, pointed out the religious ideas generally are optimally counter-intuitive insofar as they tend not to be extremely foreign to everyday experience while also not completely mundane (see also Barrett, 2008). They explain such consistencies on the basis of universal innate domain-specific processing mechanisms. An enactive account would drop the necessity of domain-specificity programmed during the Pleistocene era and look toward the kind of sociolinguistic features of current and past life to see what makes such psychologies robust. Whenever humans are in a new environment, they tend to self-organize in a way that maintains such patterns across history.

Future directions

The purpose of this chapter was to harness previous chapters in a direction that can expand CSR into new realms that account for the way humans shape the realities they experience. I summarized CSR to illuminate the ways in which it draws upon evolutionary psychology. The problem with relying on evolutionary psychology is that it only includes culture as a phylogenetic feature of the environment that is part of natural selection. It bypasses religious phenomena by simply defining it as theoretically irrelevant. I attempted to substantiate how the ontogenetic role of culture is bypassed and that rectifying this bypass is necessary to expand CSR in a direction that can adequately account for religious phenomena. My proposal rests upon a strong form of cultural psychology that highlights how culture and cognition are irreducibly interdependent by way of sociolinguistic practices. Apprehending this interdependence is crucial for addressing culturally shaped meaning and experience that constitutes the phenomenologically immediate realities. As a way to establish a more robust CSR, I proposed an enactive approach. By providing an alternative paradigm of cognition that accounts for phylogenetic and a meaningfully ontogenetic role of culture, I hope to offer new directions in CSR that can enable researchers to work within the realities of their religious participants and still have a theory of cognition.

In Chapter 1, I spelled out three criteria for truth that will enable us to realize a Jamesian approach to religious belief. Chapter 2 addressed the notion that truth should fit with life. Turning to cultural psychology allows this criterion to be realized because we can conduct studies that apprehend the purview of life lived by research participants. Chapter 3 discussed how truth also fits with a community. I took the argument a step further by providing an approach to cognition that fits with researchers in CSR and/or researchers in cultural psychology. Such work paved a way for the discussion in this chapter, which is that truth is generative. Extending the previous chapter with this one allows for us to move forward in new ways that include religious experience. In the next chapter, I make a transition to discuss what happens if we take the foregoing seriously. That is, I turn to the notion of what counts as *good research.*

Chapter 5

On being good researchers[1]

Abstract

Previous chapters involved an implicit motivation: a commitment to good research. Good research usually refers to research that is technically done well according to the methodological formulae, and moral "goodness" is usually relegated to an ethical review process. I review several scholars that challenge this common practice and argue that it is impossible to separate personal moral life from professional research practice. As such, this chapter presents how researchers in CSR ought to become broader in terms of what good research entails. This presentation involves the expansion of the notion of "goodness" to include virtue: good research is virtuous research. I first recap James' approach to discuss how it aligns with the notion of virtue. The chapter, then, via illustrations of current work, discusses how researchers in the cognitive science of religion want to be value neutral in their quest for a scientific psychology but simultaneously espouse a virtue of methodological rigor – an unstated proscriptive goodness. I challenge this ethic as the sole virtue and raise a discussion of good in terms of the character of one doing the cognitive science of religion. Such a discussion rests upon the importance of careful interpretation and coming to personally understand a community.

> But the world each of us feels most intimately at home with is that of beings with histories that play into our history, whom we can help in their vicissitudes even as they help us in ours.
>
> (James, 1912, p. 49)

Introduction

I recently corresponded with a CSR researcher named Laird Edman who draws upon the notion that retention and transmission of religious concepts is due to the constraints of human cognitive architecture (Edman, 2014, 2015; Edman et al., 2015; McConnel & Edman, 2013; see Barrett, 2008; Boyer & Ramble, 2001). The idea is that our cognitive structures entail concepts that can be intuitive, and these concepts are the usual run-of-the-mill concepts that we deal with in everyday life. For example, when an object is not suspended by something, it falls, and we intuitively find this sensible. Edman is interested in *minimally* counterintuitive concepts (MCI) that defy one or very few intuitive expectations. Such a concept would be something like the ghost of an ancestor that has all the same qualities of a normal person except that it has no corporeal body – it violated one intuitive expectation about agents. Maximally counterintuitive concepts (MXCI) are concepts that substantially violate intuitive expectations (see Barrett, 2008). Edman draws upon the idea that religious concepts that are most likely retained and transmitted when they are MCI. His work investigates the effects of counterintuitiveness on memory and reaction time in reference to theological concepts. Using stories, he studied the effects of level of counterintuitiveness of complex abstract Christian theological concepts on how long it took participants to recognize concepts (i.e., intuitive concepts would be recognized quicker) and recall (i.e. intuitive concepts would be easier to recall). He was interested in applying and connecting CSR with Christian ideas and so attempts to practically speak to a particular religious community.

Edman's work can serve as an illustration of what emerges from the previous chapters by outlining what is implicit in his work: a commitment to good research. A challenge is that *good* research is a tricky endeavor because, as the epigraph above indicates, people feel at home in what they know, and this can provide blinders that limit our research. Usually, good research refers to research that is technically done well according to the methodological formulae (Billig, 1996; Gergen, 1985). Stam (2015a) notes that this presupposition emerges from an implicit distinction that researchers tend to make between the realm of research ethics and the conduct of research. Researchers submit a project for ethical review at a university research ethics board to make sure that the project will not harm anyone. The notion of "goodness" in terms of morality is relegated to the ethical review process, and the result is that "good" research often involves standards of methodological rigor that don't violate bureaucratic procedures defined by research ethics boards. Several scholars have challenged the separation of moral goodness from research practice (Cresswell & Baerveldt, 2016; Teo, 2015; Walsh, 2015). Walsh (2015), for example, points out that ethics are not a peripheral part of our world because all aspects of life are profoundly moral – including the act of research. As I will discuss below, such claims point out that the personal commitment to methodological rigor is itself a kind of moral claim, and moral goodness is still smuggled back in (for extensive discussion see Fowers, 2015). It is not possible to separate personal life from professional life insofar as moral goodness is integral to life. A result is that researchers like Edman are actually steeped in moral goodness even though there is an ostensive split between research ethics and praxis. As such, a discussion about ethics and the role researchers play in society is the focus of this chapter (see Sugarman, 2015; Teo, 2015). An important quality of this discussion is how research that relies on the bifurcation addressed above can result in a climate where researchers miss their responsibility to others in general and those they conduct research with in particular. This chapter is a discussion about how researchers ought to become more ideologically aware of what good research entails and Edman's work will serve as an illustration for clarifying this discussion. Key to this discussion is the expansion of the notion of "goodness" to explicitly encompass the notion of virtue and go beyond goodness in terms of just research methods or ethical codes of conduct. That is, good research is virtuous research, and this chapter will address the practice thereof. I will first recap a few ideas about James' approach and then discuss how it aligns with the notion of virtue.

Good religious research: James' style

I started this book with a discussion of religious belief and how James argues we ought to approach the study of belief. James was critical of an abstract approach to belief that involves a separation of something from a belief about the thing. He was against the idea we can sensibly talk about belief being separate from something in the way that an abstract and disembodied approach implies. His work informs us that we must address what beliefs are *about* if we plan to study religious belief and so turn to a cultural approach that addresses the narratives and stories that include the content of belief (see Chapter 2). The foregoing chapters attempted to articulate an approach that meaningfully includes experience. Experience involves the body and so addressing this question involves a discussion of the body (see Thompson, 2007). James goes so far as to claim that belief is best characterized in terms of pure experience, which is the name for a collection of "sensible natures" (1996/1912, p. 27). Belief is entwined with our very sensation of the world around us because, as James (1982/1902) notes, thoughts or like features of our psychology have compelling force that has ontological gravitas.

Through the argument presented in Chapters 2 through 4, I have advocated a turn to realities as people experience them instead of realities that researchers determine a priori. A revamped CSR is one that looks to what people experience as real insofar as what "these people experience IS reality. It gives us [and them] an absolute phase of the universe" (James, 1981/1907, p. 17; original emphasis). It may seem like this pluralist positioning disconnects us from any form of objective science, and this charge is partially true. When we take an approach outlined by James, we turn to the shared realities that are not entirely objective. Such realities, however, are not subjective either. When I addressed criteria for truth claims in CSR, I addressed how they must involve a fit with the life of those we study. It is for this reason that we do not collapse into subjective fancy. James (1981/1907, 2011/1909) repeatedly argued that truth is not a matter of the mere proclivities of subjectivity. He did so in a way that points out how one's truth is only true when it resonates with life outside one's mind and so means that it resonates with others.

He made the claim that standards of fit are socio-normative in quality and are integrally integrated with life with others. This claim is not to say that people are bullied into experience by authority, but rather that those around us shape what we experience as real in the ways we outlined in previous chapters. As we come to new experience, our repertoire of facts grows with our experience as new truths are enfolded into current experience.

I propose turning to the facts of experience as people live them and so researchers cannot make any sort of fanciful claim. As such, it is possible to handle religious experience while staying grounded in facts (for a detailed discussion see Chirkov, 2016). Edman's project has leanings in this direction insofar as he turns to the theological concepts pertaining to a Christian community when he approaches religious beliefs and ideas that are part of a reality people live. Where he departs from this criterion is in his commitment to MCI and MXCI as features of cognitive architecture that he believes in a priori. In one way, he engages people's realities by discussing ideas pertaining to a community and in another, bypasses it at the same time by not explicitly addressing the perspective members of the community's take on his work.

There is an important ethical implication bound up with understanding facts taken as ostensive reality within a community. It has been described by Teo (2008, 2015) as "epistemic violence" that emerges from the speculative nature of psychologically oriented research. That is, for example, we do not literally see MCIs or XMCIs but substantiate their ontological status through research presuppositions furnished by our respective paradigms. Much of what we write about in CSR in general is strictly speculative because the subject matter in cognitive science cannot literally be seen. Teo argues that speculation is integral to many efforts in the psychological sciences because the objects of our inquiry cannot be accessed in the same way as physical sciences. Even though researchers may do their best to see their work as non-speculative, it is an essential component of research:

> Although interpretations of results are presented as facts, they contain speculative elements. As suggested above, *knowledge* that is produced in empirical psychological studies contains data *and* interpretations. Interpretations are not determined by data and thus require a hermeneutic process.
>
> (Teo, 2008, p. 56; original emphasis)

Teo is echoing a wide range of theorists who have noted how researchers define and determine results as a function of their paradigms as data do not speak for themselves (Danziger, 1990; Gergen, 1985; Stam, 2015b). They echo a claim in the style of James that what counts as facts of experience is shaped by communities of practice.

The speculative quality of research is not a problem so long as researchers are content to talk among themselves. That is, the speculative quality of CSR is not a problem so long as researchers are not intending to say anything to religious believers because there are few problems or dangers when one speaks from within a community of practice. The challenge is that researchers cannot remain cloistered in ivory towers and researchers like Edman are commonly branching out to speak to practical concerns. Such action is good, and more granting agencies are calling for articulation of "real world" implications. It is in this context that Teo (2008) notes how speculation has serious practical, behavioral, or existential consequences, and I argue that such consequences are rooted in the manner that speculation is a communally situated practice. One community can relegate another's purview as inferior and

> as soon as these speculations construct the 'Other' as problematic or as inferior, with possible negative consequences for the 'Other,' one should speak of a form of violence that is produced in 'knowledge.' In these cases, interpretative speculations (and not data!) turn into *epistemological violence*.
>
> (Teo, 2008, p. 57; original emphasis)

There are extremes in such practice such as Dawkins' well-known supposition that people who are religious could be constituted as belonging to the "dulls" and those who are atheists relegated to the clear and rational thinking "brights." Such a speculative move clearly denotes some people as inferior to others on the basis of religious affiliation. The consequences are blatantly violent and oppressive wherein one community of practice denigrates another. Teo would see such actions as clearly problematic, but what about Edman and his work to understand MCIs in a Christian context?

In no way is Edman's work *intended* as being antagonistic or oppressive, for example. He reflects a general position in CSR to be kindly agnostic if not subtly supportive of religious belief (c.f. Barrett, 2004, 2012). Stam (2015a) admonishes researchers to consider how psychological language practically impacts those that we study and crystalizes a challenge to CSR. He notes that researchers must be competent practitioners in the communities of science in which they circulate. Claims made by researchers among themselves are subject to an ethics of shared understanding insofar as they must comply to norms and standards of communal existence. A challenge is that psychological terminology has leaked into public discourse and circulates widely. It is common for people to talk about self-esteem or self-efficacy, for example. Stam notes that a scientific veneer accompanies language emerging from researchers, and with that veneer comes implicit authority and power. Researchers obtain status as experts on people in a way that establishes a kind of authority over others. The danger in CSR is that it is possible to commit a form of epistemic violence when we claim to know someone else. That is, the foregoing chapters wherein I spell out how researchers bypass or reduce religious belief to a "nothing-but . . ." is an articulation of how epistemic violence can be perpetrated on others. Edman is not immune to this danger insofar as his presuppositions can simply define what matters to religious believers out of relevance. It can potentially shut down others by problematizing their own self-knowledge, and this amounts to a subtler form of epistemic violence.

This sort of argument is what leads to the necessity to move the discussion of ethics beyond that of prescriptive codes of research ethics boards to that of moral sensibility (see Stam, 2015a; Teo, 2015). Often, ethical codes miss the kinds of human consequences that need to be considered because they pertain to the community of researchers and do not extend to the communities in which researchers are working. For example, would Edman's research impact the Christian community he is interested in helping in ways that are outside the scope of research ethics boards? Teo argues that an orientation to such questions means that

> problematic issues are addressed within moral-practical inquiries. Solutions in the moral domain demand a break from seemingly self-evident actions that we engage in, for instance, as psychologists. Applied to psychological ethics the following question could be asked: "What does it mean to do the right thing as a psychologist?" Answering this question requires intersubjectivity and the involvement of the *Other* (student, client, patient, participant, etc.).
>
> (2015, p. 79)

Despite writing to psychologists, Teo's comments apply CSR insofar as a more robust science would adequately account for the religious beliefs and experiences requires a discussion of moral practices.

A more robust CSR requires researchers like Edman to pay closer attention to the way that humans establish the facts of experience in a moral sense. James argues that the facts people take to be true are made in the course of experience. That is, "[i]t is the most personal experience of those most qualified in all our circle of knowledge to HAVE experience, to tell us WHAT is" (James, 1981/1907, p. 17). When I discussed that truth of religious claims can be substantiated by virtue of their fit in life, I was addressing the importance of the experience people have in everyday life. James (2011/1909) expanded the definition of truth by arguing that it is always inextricably bound to life. Truth is not used in a moral vacuum because people use it in relation to all of the other aspects of life. He links this kind of effort to the notion of moral goods and argued that facts of experience are such when they fall into rhythm with life and reflect satisfactory relationships with other parts of experience. This satisfactory quality involves facts that are a *good* fit with the rest of life. The pluralism inherent in James means that it is not a moral system of rules that we are discussing here. It is about the good that fits with life in the sense of good behavior. Goodness involves resonance with normativity of a community and this notion cannot be separated from moral goodness. Facts are whatever prove themselves to be good insofar as they fit with life and do not violate morals constituted in lived normativity.

Truth claims made by researchers or anyone else are common-sense claims, and it means that a set of facts will fit morally in one sphere of life but perhaps, not another. James points out that such sociocultural constituted facts continue on something like a "credit system" (1981/1907, p. 85) where people take them for granted until they are ruptured. For example, in Edman's work, he discusses different theological concepts, and we can imagine how different denominations in Christianity would have different concepts that are taken for granted. When one comes into contact with different concepts, then the facts of experience can be questioned, and Edman's work depends upon this happening. He expects some foreign religious concepts to be experienced as more counter-intuitive than closely related ones. Edman implicitly relies upon the fit with life in his research design and presuppositions, but he misses how there is a moral dimension at stake: some concepts are morally right in a community.

What turning to a more robust CSR allows is for us to develop a more sophisticated approach to such competing moral ontologies and things like God's existence. I discussed how truth can be ascertained through considering whether or not a claim is generative. We change and the world around us changes such that a true belief could become antiquated and irrelevant. A true belief, James (1981/1907, 2011/1909) argues, is one that fits the dynamics of life to do more. Consider how he argues that there is no meaning in denying God's existence because the idea of God is itself pragmatically useful in some communities. It works and fits the facts of experience for many people that we seek to study and so proceeding as if it is true moves us forward in understanding others. If an idea has value insofar as it works in life, it has enough to be a fact of experience, which opens new directions. We are not wed to preconceived ontological notions, and we still can study religious experience without falling into reductive approaches that enable epistemological violence.

A challenge is that we cannot just collect information about various groups. When a dispute over the facts of experience becomes serious, the facts themselves can be questioned, and ultimately one side of a dispute must show more practical difference over another, but this is not enough. If we want to move Edman's work in a more robust direction, it would involve questioning all facts, including those presupposed about MCI/MXCI on moral grounds and whether or not there is epistemological violence at stake. It would involve questioning implicit ideas and claims that are essential in the practice of Edman's appropriation of CSR and in the life of religious believers. Conversely, it includes drawing on MCI and MXCI to challenge and question religious belief.

The forgoing can respond to a call to researchers to realize their emancipatory potential in the face of injustice and oppression (c.f. Walsh, 2015; Walsh & Gokani, 2014). Walsh (2015) has written on the topic of ethics in the context of psychological research, and he notes that researchers ought to develop a critical consciousness of socio-contextual phenomena. This claim extends to a more interdisciplinary practice of CSR insofar as research practices can be entwined with good citizenry where dominant practices in a community can be challenged and exposed through a kind of juxtaposition. In an earlier paper, he and his colleague challenge a taboo in psychology about addressing social privilege and such an address emerges in the recognition that all aspects of our vocation are profoundly ethical (c.f. Walsh & Gokani, 2014). Such works highlight how work in CSR lacks a moral compass for addressing issues such as oppression and violence. It also highlights how there is potential for researchers to expose practices that are oppressive and propose emancipatory alternatives.

Overall, Edman's work is unique precisely because he is trying to bring about change via the use of CSR in Christian communities. He seeks to make pragmatic change in a Christian community, but he does not turn back on CSR or engage in questions about the moral goodness of his work. Edman's is trying to do good research that encourages human flourishing in the community of religious believers. That is, he is attempting to engage in a project that identifies something good, but he misses a discussion of what his work actually does to help the Christian community or CSR flourish – i.e., encourage the best in people (see Fowers, 2015). It is this notion of flourishing that brings us to the issue of virtue.

Virtue

Fowers' (2005) work addresses the notion of virtue in relation to psychology in a way that can be extended to how we can improve CSR (see also Richardson, 2012). He notes that the discipline of psychology is often concerned with science, and it focuses on making this concern core in pedagogical and scholarly practice (see for example Halonen et al., 2013). The kind of science in psychology generally models itself after the hard sciences by presupposing that science reveals a-temporal laws of human behavior – natural laws by which the human operates (Stam, 2015a). Fowers and others have identified that this preoccupation with universal covering laws is an unsubstantiated assumption in psychology (see Bishop, 2007). Really, psychologists are concerned with an idealized version of science that itself amounts to an ideological moral imperative (Brown & Stenner, 2009; Danziger, 1990; Harré, 2002; Stam, 2015a; Teo, 2008, 2015). In my conversations with Edman and others in CSR, I have come to realize that scientists in CSR are often *personally* concerned with a similar need to use sound empirical research methods. Edman is concerned with studying MCI and MXCI using experimental methods aimed at producing *good* scientific research. The notion of "good" highlights how the practice of CSR includes a kind of virtue where *good* work is work that falls in line with this idealized vision of science. Fowers illuminates how, on the one hand, Edman and other researchers in CSR want to be value neutral in their quest for a scientific psychology. On the other hand, they simultaneously espouse a kind of value system that amounts to proscriptive goodness.

Furthermore, moral injunctions show up in psychology when we consider how psychologists conceive the purpose or ends of their work. Many national psychological associations include a mission statement to the effect that scientific efforts should enable people to be functional, healthy, and effective (see for example Halonen et al., 2013; c.f. Sugarman, 2015). What this means in terms of Edman and CSR is that, on the one hand, there is a moral injunction for ethical neutrality as good scientific practice while, on the other hand, there is a value system espousing that *good* work ought to have an individualistic and instrumental value – improve the personal quality of life for people. Edman's work is explicitly about improving the work of ministers in church settings, but the challenge is that he could commit epistemic violence without adequate reflexivity directed to his own prejudice. Turning to the notion of virtue helps us address a kind of moral paradox in CSR, which also makes its way into Edman's work.

Consider how Fowers (2005) points out that researchers tend to betray their own objective of ethical neutrality with their values about what is good research and the good outcome of research. Fowers is predicating his discussion on one that discusses two questions: Can social science research be neutral? Should social science be neutral? Bishop (2007) explores these two questions in greater depth by discussing how the answer to both is "no." An objective researcher does not need to pretend to be value neutral because it is disingenuous. Many of the most persistent findings in science are dependent on values and meanings that are pervasive at the time. He points out that being value-neutral is, paradoxically, a value of political liberalism. This is an example of a disguised ideology that underlies attempts at value-neutrality in research, and it overlooks important issues like moral obligations to others (e.g., other scientists) and the importance of human agency. That is, political liberalism undergirds research and so a sort of virtue is already implicitly integral to science and CSR. The problem is not that researchers include a conception of goodness in their work, but rather that it is not honestly acknowledged. Human efforts are never free of "moral color," says Fowers, and so it is prudent to make explicit the presuppositions that are implicit:

> the adoption of a virtue framework would not represent a shift from a value-neutral discipline to a psychology that included a moral dimension. Rather, the incorporation of virtue in psychology represents a shift from an implicit and unsystematic approach to moral matters to an explicit and well-formulated framework, an openly articulated framework that can be questioned, debated, and evaluated.
>
> (p. 22)

Doing so allows us to expand research in CSR and get at the kinds of human phenomena that researchers tend to bypass. Separating facts from values is problematic because it is artificial as interrelated goods always emerge in human affairs. Even at a general level, the pursuit of knowledge and freedom of inquiry that is valued in the human and hard sciences is a value and brings us back to moral inquiry. Separating science and fact from notions of value and the good are doomed to create the kind of bypass that CSR has committed. The project of this book has been to propose a new approach to the CSR that brings implicit virtues to the forefront and into tension with others.

Fowers (2005) notes that virtues cannot be understood in piecemeal fashion or through formulaic empirical measurement and that codification of ethics is insufficient (see also Teo, 2015; Walsh, 2015). When it comes to what is good, we can't predetermine what will be good in every situation like methodological prescriptiveness and research ethics boards require. Implicitly venerating experiments through adherence to methodological prescriptiveness circumscribes how we ought to do our work and shuts down the dynamic relationality required for understanding others' realities and avoiding epistemological violence. Fowers (2005), for instance, discusses psychological science in terms of techné, which concerns the technical proficiency of an actor to execute actions that are needed to bring about good work. Good techné in CSR would normally be the ability to carefully construct and implement an experimental design. Edman, for example, designed careful studies to measure and assess MCI and MXCI by presenting participants with various concepts and systematically measuring how they were rated. A challenge is that technical quality of the product has nothing to say about the product or its place in life. A focus on techné detaches our work from ethical implications even though it smuggles in virtue. In CSR, techné alone doesn't work because we need to be concerned with the quality of goods we provide and what they mean to a community. Edman's work is concerned with how it speaks to religious communities and of what worth it is outside of the research laboratory. It is in this way that he illustrates more than a simple study with no concern for human flourishing. Fowers, however, inspires the idea that we need to redefine CSR in ethical terms instead of just techné. Critique of Edman's own efforts ought to go beyond techné to the ontological and moral presuppositions that substantiate it.

Turning to virtue allows us to approach the kinds of phenomena that people live in everyday life and so resonates with a Jamesian approach. Every day, people dwell in the kinds of qualities that are described by virtues. Consider how we view issues like moral character when determining our friends and lovers (see Richardson, 2012). It is a strange oversight how researchers in CSR overlook the importance of personal excellence when it is what we most admire in persons. That is, we often think about the kind of person we want to connect with but do not do so when it comes to considering what kind of researcher one ought to be. Life often involves higher-order goods that we look for in others insofar as we must work out higher-order goals that guide our lives. Fowers (2005) notes that, where psychological theory often sees personal subjective well-being as the highest good, there is more to this discussion and the kinds of qualities that bring out the best in us. It is important to extend this notion to the kind of person religious believers would want doing research with them. We are brought to a place where "the legitimacy of contemporary science is seen largely in how it serves ordinary lives" (ibid., p. 163). We cannot explain or address human behavior without turning to virtue because it is integral to efforts, whether as researchers or the researched. Whether in religious life, non-religious life, and/or as academic researchers, we fall back into a notion of what is good: virtue. To understand religious realities of the sort outlined in previous chapters, we need to consider what others consider virtuous. That is, for Edman to study MCI and MXCI concepts, he first started with something that is *good* to address within the Christian tradition: Christian theological concepts. He is pursuing goods, but a fuller approach to CSR would entail the moral character entailed in CSR. Far from approaching dogmatism,

I am advocating for the juxtaposition of realities to ask what is good in terms of the character of one doing CSR.

The previous chapters highlight how much of what we take to be mental properties are culturally constituted phenomena, and virtue, argues Fowers (2005), is no exception. He argues that virtue involves communally located participation in shared goods that grows out of their social nature. The most precious goods can only be held in common insofar as the kinds of things we think are very good tend to be the kinds of things that others also think are good: "One can attain internal goods only by acting in the ways that embody those good. One pursues internal goods by participating in the activities that constitute these goods" (p. 65). Goods are lived in communities of mutual participation such that what is good for me is good for us. Richardson (2012) draws upon Fowers to argue that central to virtue ethics is dialogical understanding and our dependence upon others to learn what counts as good. A strong form of relationality is needed because our ability to act virtuously involves sensitivity to others. It is for this reason that the pursuit of goods is dependent on others and virtue is seen in communal patterns of practice. Pursuit of the good does not subordinate individual aims to the good of a community in some oppressive form. Rather, shared goods provide higher-order ends that provide the context for individual striving, which means that flourishing is deeply dependent on participation in communities.

Such an approach raises a challenge against the individualist presuppositions and subtle epistemological violence that can potentially emerge in CSR. Fowers argues that a virtue approach is "more appropriate for a subject matter that has strong context dependency and is concerned with understanding humans as beings whose choices and actions are guided by what they see as worthwhile and good" (2005, p. 154). This means that, to practically understand others, one needs to think in terms of a deeply pervasive form of context dependency. It means to apprehend the realities in which others live, but the problem is that CSR often works within a paradigm wherein individuals are prior to society. For example, Edman's work is predicated on the importance of cognitive architecture that enables MCI as opposed to MXCI to be retained and transmitted over generations. One the one hand, he attempts to get at something relevant to those within Christian communities. On the other hand, he does so without really grasping the culturally constituted realities that such people live, which leads to missing the contextual dependency of religious cognition on culture and epistemic violence. If he were to look back at his own reality, he could see how his presupposed realities are shaped in a community of practice committed to CSR that precludes a deep sense of the socially constituted realities of his participants. This example reflects how there is a rich social world that CSR neglects and so cuts itself off from what is virtuous within the community. Shared goods are virtually invisible to Edman and CSR due to the individualist focus.

Fowers (2005) argues that making visible shared goods involves research modes that meaningfully include *phronesis*, which he draws from Aristotle. Phronesis is best understood as practical wisdom or judgment, and it amounts to the capacity to make and apply good judgments in everyday life. Research that meaningfully includes phronesis is research that involves researchers apprehending the everyday life and world views of others in a way that would preclude epistemological violence. A researcher who takes phronesis seriously cultivates the ability to see what is important to the members of a community being studied. It includes understanding how to make wise choices per what that community sees as good and act on the choices in such a way that one is seen as acting with goodness within the community.

Fowers rests this suggestion on the hermeneutic tradition inspired by authors such a Dilthey and currently manifest in contemporary thinkers such as Gadamer (2008) and Taylor (1999). Bishop (2007) condenses this work into a summary that highlights how researchers are part of what they are studying. This position means that they become familiar with the life as those they study see it. Hermeneutic *interpretation* means that there is no distinction between a person and the world, which means that coming to personally understand a community means coming into an understanding of something that it not reducible to one's own subjective proclivities. Just as dialogue in everyday life brings people to a place of understanding others, dialogue with participants enables a sense of their life and so sets the ground for insight. The results are a moral understanding that is about what practically works for others in a community and treating participants as they take themselves to be and not in terms of some a priori ideal. In the case of Edman, interpretation would involve being able to make good judgments in the everyday life of the religious communities he is interested in studying. His interpretive work would entail phronesis if members of the Christian communities in which he was studying saw his work as good for them. What a consideration of virtue adds to Edman is a deeply context-dependent moral perception in his research. He needs to explicitly and meaningfully incorporate what is essential and what is practically good in the communities he is studying to embody phronesis. His interpretive work would demonstrate how he could distinguish what matters in the community from what is trivial by knowing what terms are important and how MCI and MXCI are not trivial.

According to Fowers (2005), a turn to virtue would be a turn to phronesis, which means that interpretive work in CSR would amount to perception of what is at stake within a community, deliberation about possible actions within such a community, and reasoned choice about the best course of actions within the realities lived by religious practitioners. A challenge for researchers, like Edman, is that phronetic interpretation is not a matter of technical expertise. Technical execution of research methods in the natural sciences precludes the role of judgment and familiarity, which are essential if we are to include phronesis. Instead of focusing on the technically appropriate execution of research, one focuses on a researcher's judgments about the good one seeks for the community of interest. There is no technical formula for this kind of work because it simply entails getting involved with the community to apprehend the realities that they live. Edman implicitly relies upon familiarity with his own experience as being part of a Christian community, but still reverts to the technical execution of research rather than drawing upon his own experience with the community. When Fowers (2005) argues that we engage in deliberation as researchers, he argues that judgment is not always clear because there is no set of universal rules or unassailable methods that we can rely upon with certainty. Deliberative interpretation about the virtuous course of action involves thinking from within a community about the most relevant of multiple complex components.

A central facet of this discussion is making explicit the virtues researchers enact as part of a community. In addition to techné, argues Fowers (2005), researchers have a kind of virtue that values individual subjective well-being and the personal achievement of happiness (see Sugarman, 2015). Discussing virtue raises CSR beyond mere subjective well-being to how to contribute to human flourishing. Juxtapositioning realities enables discussion about the kinds of values that will enable people to flourish and investigating what kind of habitual excellence in action contributes to this aim. That is, it is possible to raise questions about how researchers characterize themselves and the kinds of character qualities we see in researchers relative to those we are studying. Virtue is about developing human ethical potential for more than just individuals, and I am seeking to provoke CSR to consider what is good for many groups: researchers developing scientific knowledge while being able to speak to religious communities and vice versa,

Virtue, Fowers (2005) argues, does not provide a set of rules but uses the concept of goodness and virtue to guide actions. This position frees us from any formal or necessary relationship among religion, politics, and virtue as turning to virtue is about how to live a good life and not how to sit in judgment of others. What is good is always open to debate as one can't know what is best for all people in a conclusive way. It is such a discussion that enables emancipatory potential in research. That is, standards of excellence and goods change with communities, and the foregoing enables a deeper discussion about goods in different communities relative to each other. Critique becomes intrinsic following the due diligence needed to avoid epistemic violence (see Cresswell & Baerveldt, 2016).

Richardson (2012) expands on the notion of phronesis to argue that incorporating the notion of *the good* and the examination of character is difficult because researchers do not want to violate their positions as scientists. Researchers in CSR, however, must still retain enough authority or credibility to be seen as legitimate in communities like other social scientists. In order to reconcile this tension, Fowers (2005, 2015) and Richardson (2012) both argue that research should be less concerned with finding definitive answers and more concerned with generating good questions. Problem finding is a mark of the best scientists who take phronesis seriously. Good questions are a matter of moral perception insofar as they often involve goods as they are understood in a community. Researchers in CSR need moral perception to recognize a problem and pursue it. Rather than judging the quality of investigations by the technical correctness of the research method and design, the quality depends upon the quality of the questions asked. Consider the example of Edman's research (e.g., Edman et al., 2015). He did not find clear-cut results insofar as he could not statistically identify what theological concepts were MCI and what ones were MXCI or what differences reliably predicted. Unfortunately, spotty results are not often celebrated, and his study was largely unpublishable because of it. He did find that theological concepts are not easy to map onto the MCI and MXCI theoretical paradigm. The very messiness of his results is what shows he is onto a real problem and opens the door for an inquiry into what is good within a community. The mixed results are a good thing because they open up the potentiality for deeper questions.

The foregoing discussion on virtue spells out what kind of research would constitute good research, and it also resonates with James' approach to religious belief. James inspires an approach to religious belief that treats it as embodied and not abstracted. An approach oriented toward virtue challenges the preoccupation with universal covering laws. When James advocates a turn to the concrete specifics of religious belief, he is also challenging the same abstract approach in CSR. Expanding research in CSR gets at human phenomena as they are lived and resonates significantly with a concern with virtue that emphasized the lived practices of everyday life. James works especially hard at articulating how what constitutes truth is what is good as part of a community. James and Fowers align with each other in their shared claim that an idealized version of science amounts to an ideological moral imperative pertaining to a particular community: researchers. In Fower's discussion of the importance of dynamic relationality required for understanding others' virtues, he touches on the importance of experience. For instance, he discusses psychological science in terms of techné and highlights how it neglects the experienced familiarity with life. As such, when he advocates for recognition of the limitations of techné, he is advocating a turn to experience just like James. That is, his discussion of virtue emphasizes experienced realities inherent in religious belief.

Furthermore, the criterion that truth can be determined through fit with life is a Jamesian idea that resonates with virtue. Fowers (2005) brings us to the place where the legitimacy of science is seen in how it serves ordinary lives, and this claim is a recapitulation of James's concern with a fit with life. When Fowers argues that human efforts are never free of "moral color," he is arguing that our concern with any kind of truth claim is situated in a context of how good it fits with life. Expanding the notion of fitness with life brings us to the James' second criterion for truth, which is that it should fit with a community. Previous chapters highlight how much of what we take to be mental properties are culturally constituted phenomena, and virtue, argues Fowers (2005), is no exception. To understand religious realities of the sort outlined in previous chapters, we need to consider what others consider virtuous and good. This approach leads to a strong form of relationality, much like James. It is for this reason that the pursuit of goods is dependent on others and virtue is seen in communal patterns of practice, just as religious belief is a communal practice. This means that to practically understand others one needs to think in terms of a deeply pervasive form of context dependency that spells out what is true and what is good at the same time. Last, when researchers turn to an approach that considers virtue, they turn to a concern with generativity. Making visible shared goods involves research that meaningfully includes phronesis. Developing an insight by interpreting an other's community and what is at stake is to develop phronesis. A key, moreover, is to consider how having both an inside and outside apprehension of a community enables juxtaposition. Retaining an outside perspective and not only thinking like those we studied enables generativity because research should be less about finding definitive answers and more about finding good problems or potentially oppressive practices. Good questions are a matter of moral perception insofar as they often involve goods thoroughly understood from within a community while potentially questioning such goods from an outside perspective. This direction is exactly what James was interested in when he set out a truth claim as something that should be generative.

A counter example

Edman's work already illustrates some of the kind of changes I am proposing, but it is also helpful to discuss a counter-example that illuminates the kind of work I seek to provoke and critique. Gervais and Norenzayan (2012) produced an article pertaining to religious belief, resulting in significant public attention. The article, entitled "Analytic thinking promotes religious belief," found its way into the public consciousness to such a degree that a Google search on "analytic thinking" in 2012 listed a *Scientific American* piece covering Gervais and Norenzayan's article as second on the list (Krakovsky, 2012). The article became central in public discourse at the end of April 2012 when it was released (e.g., Ball, 2012) and continued to remain in media discussions (e.g., Brown, 2012). The title of Gervais and Norenzayan's article expresses what they believed to have found, the implications of which can be seen in media titles such as "Analytic thinking can promote atheism" (Pappas, 2012). The reception of this CSR study leaves readers with the impression that religious belief is inversely related to analytic thinking. My concern is that such an impression resonates with the epistemic violence we seek to avoid.

First in reference to virtue in terms of doing good research from within a CSR paradigm, Gervais and Norenzayan (2012) report on five studies that attempted to address religious belief, but the authors' conception of it is too shallow to make the claims that they make. For example: (1) study 1 addressed the salience of belief with the notion of intrinsic religiosity (e.g., depth of commitment to religious practices) and not belief per se, (2) study 2 addressed religious belief in terms of a ranking of 0–100 in response to merely one item asking if participants believe in God, and (3) studies 3 through 5 assessed religious belief with three items ranking the degree to which participants believed God exists, angels exist, and the devil exists. Consider how James would argue that such beliefs are so general that they do not adequately grasp what belief is about. While commitment is part of belief (study 1), it is more than simply motivation to participate and a ranking on how much one believes in God on a scale of 0 to 100 (study 2) tells us nothing about what God is believed to be. Ranking the degree to which one believes three kinds of entities exist commits the same oversight. None of the assessments of religious belief come close to addressing the experiential aspect of belief and how they are part of world views lived by religious practitioners. Gervais and Norenzayan make the claim that "[i]f religious belief emerges through a converging set of intuitive processes, and analytic processing can inhibit or override intuitive processing, then analytic thinking may undermine intuitive support for religious belief" (2012, p. 493). Such a quote has been picked up by the popular media, but it is potentially misleading. Others have taken the challenge of redressing some of the problems inherent in this claim (see Yonker et al., 2016). If we were to consider techné alone, these measures are arguably inappropriately shallow approximations of the well-developed systems of religions believers.

Virtue involves evaluating actions and claims in reference to standards of excellence in a community and from within the community of CSR, the article is less than admirable. Although Gervais and Norenzayan explicitly draw on CSR, the measures used therein are not really consonant with this approach either. For example, McCauley (2011) writes about maturationally natural religious beliefs, which are beliefs that are hard wired in by way of evolutionary processes. Maturationally natural religious beliefs generally pertain to opaque super agents that are much like humans except for one or two deviating features (see Barrett, 2008; Keleman, 2004). When Gervais and Norenzayan (2012) invoke the notion of "God" (including the capitalized moniker in the measures), they invoke beliefs that pertain to more technically specific religious formulations that are learned. While maturationally natural beliefs may be intuitive, they are often theologically incorrect insofar as they would not include notions such as God, devils, and angels (see Slone, 2004). In this case, the authors actually assessed religious beliefs that are acquired through learning and enculturation. Therefore, the authors measured theologically correct beliefs that can be taken to be true, yet often associated with deliberative theological endeavors. In sum, the authors sought to display the overriding of religious beliefs with analytic thought, but they did not measure religious beliefs in an appropriate manner or as they have been researched by CSR standards.

The study attempted to show how analytic thinking overrides intuitive religious beliefs by drawing upon a distinction between System One versus System Two processing. System One processing relies upon intuitive heuristics while System Two draws upon analytic, careful, reflective, and deliberate processing. The study was predicated upon the idea that prompting System Two processing inhibits intuitive religious beliefs. A significant problem with the study was that the authors equated the triggering of analytic thinking with inhibition that comes from activities such as syllogisms that require participants to inhibit their first response to a problem. That is, the article assessed inhibition and considered this analytic thinking. This definition of analytic thinking is inappropriate insofar as it misses what analytic thinking is about. Nobody would consider a Stroop test –

a test where people are asked to read a word like "red" in green lettering that requires one to inhibit the natural intuition to say "green" – a measure of analytic thinking, but it is a test that creates the same inhibition that the authors relied upon (see Yonker et al., 2016). The idea that the authors promote is that inhibition is equitable to analytical thinking because it involves a state of mind that is deliberate or reflective. This inference is unmerited. For example, one may try to slow down and deliberately reflect on what one had for breakfast a week ago, but this effort could hardly be considered analytic thinking. The implications drawn in the public discourse are potentially unfounded because Gervais and Norenzayan (2012) do not consider analytic thinking in a manner similar to those addressed in the popular media that discuss them. When evaluating Gervais and Norenzayan according to the standards of a community, their work emerges as something other than virtuous.

Second, in terms of virtue and developing phronesis pertaining to those researchers study, Gervais and Norenzayan (2012) also fall short. At the center of the notion of virtue is the idea of phronesis, and I described it as practical wisdom or judgment and it amounts to the capacity to make good and apply good judgments in everyday life. The good judgments that people must often make are about what is good for them and reflect higher-order goods in our lives. To live well is to know how to get along with others and excel in living a life that contributes to the goals one has about how to be good. From the perspective of religious believers, the approach to religious belief in their article would not resonate with the foregoing discussion in the rest of this book. Measuring belief in terms of whether or not one believes in God, angels, or demons without any meaningful consideration of what these notions mean is an example of abstracted belief. There is no consideration of the experience of these beliefs and what they mean personally. The authors do nothing to consider if their claims would fit with the lives of anyone other than their own community of researchers, and there is little to consider whether their work has any generative value in any community other than that of themselves. The article may not fit with the cultural approach outlined in the foregoing insofar as religious belief is bypassed. A concern may be to wonder why such a bypass matters aside from not capturing the richness of the phenomena, which could arguably be an issue for researchers and few people outside that realm. With this illustration, I hope to illustrate why virtue matters and how Gervais and Norenzayan depart from the good, eroding excellence in practice. I will make this case by returning to the notion of phronesis and how generativity is undercut.

Of course, cultural psychology and James highlight how such goals may feel personal while they are, in fact, socioculturally constituted. Living well is understood in terms of communally located participation in shared virtues that grow out of social nature. Understanding what is good means to carefully interpret others to see standards of excellence experienced and lived by members of a community – an understanding and our dependence upon others to learn what counts as good. A turn to phronetic interpretation amounts to perception of what is at stake within a community, deliberation about possible actions within such a community, and reasoned choice about the best course of actions within the realities lived by religious practitioners. Gervais and Norenzayan engage in nothing that looks like phronesis insofar as there is no evidence in their paper or in the supplementary materials that offers any indication that they were interested in the communally constituted good lived by any community.

Their paper was published in the journal *Science*, which is among the highest-impact scientific journals in the world. To publish in this journal, Gervais and Norenzayan must enact an idealized version of science practiced as a hard science as outlined above. The problem is that the idealized version of science they employed cuts off the potential of phronesis of any sort other than the idealized version of science with which they are concerned. Paradoxically, they excelled in embodying the moral imperative of dressing up their work as if it were value neutral. There is still a particular value system at play in this article that embodies a kind of proscriptive methodological goodness. The tacit pursuit of this kind of virtuous research is the very thing that precluded them from being able to develop the moral perception needed to recognize what their work would mean to participants in religious or atheist communities.

In fact, the moral pursuit of supposed value neutrality undercuts any concern with the phronesis of any other communities and so allows Gervais and Norenzayan (2012) to sidestep any discussion about the moral implications of their work. The claim that analytic thinking can cause disbelief matters to people and it is for this reason that the article was picked up by popular media. Clinging only to one value system that venerates experiments shuts down the dynamic discourse that situated work in other communities. Gervais and Norenzayan's paper reflects a concern for techné and bypasses any discussion relevant to any community, and in so doing, they remain silent about the pejorative implications in their work. The challenge that considering the notion of virtue should raise is that the technical quality of the product has nothing to say about the product or its place in people's lives. Omitting this discussion leads to epistemic violence as religious believers are tacitly defined as inferior. The place of this research in people's lives, however, is exactly what matters about this study and ought to have been discussed regardless of whether the claim is true or not. It matters that religious people could be considered less analytic and simply put, dumber than other non-believers. The issue at hand is not that this claim potentially undermines the intellectual quality of religious mentality. It is, rather, that pretending to be value-neutral shuts down the potential for the kind of discussion that matters to people and is an abdication of responsibility for one's actions. We cannot explain or address human behavior without turning to virtue because it is integral to efforts whether as researchers or the researched.

Conclusion

Throughout this chapter, I discussed the work of Edman to illustrate what emerges from the previous chapters. I tried to illuminate what a commitment to good research is and how it is a tricky endeavor. This argument is intended to move us way from goodness merely in terms of research that is technically done well and raise the discussion beyond the domain of research ethics boards. My goal is to drive home the fact that ethics are not a peripheral part of our research because all aspects of life are profoundly moral, and this includes the personal commitment to methodological rigor often seen in researchers. I sought to preclude the separation of personal life from professional life to point out that we, like Edman, are actually steeped in moral goodness. I thereby offered a starting place for researchers to become more ideologically aware of what good research entails. I hoped to raise the issue of how good research is virtuous research and such a claim fits well with the Jamesian approach outlined above.

Note

1 I thank Laird Edman for his comments on this chapter.

Chapter 6

Does reality and/or God exist . . . or not?

Abstract

Discussions about religion and the veracity of the cognitive science of religion often dance around the issues of whether one believes in God (or not) and/or in the value of science. Where a religious believer may dogmatically adhere to supposed characteristics of God, scientific efforts may be hampered by near-idolatrous worship of scientific methods. This chapter is about how letting go of both modes of certainty leads to a more robust science and/or expression of religious belief. A dogmatically certain belief in the absolute in religion cuts off understanding of faith just like scientific hubris undermines knowledge in the cognitive science of religion. I review James' discussion of how we live in a pluralistic universe, which is a universe without certain appeal to the absolute and one of competing realities. This discussion enables a vision for research eschewing scientific certainty. I argue for an approach that is about understanding and transforms existence of the person who becomes humble and open to apprehending the other. A principle that guides such endeavors is thereby not a search for propositions conformable to facts, be they religious or scientific. It is the principle of charity, which one finds in many religious traditions and in good science.

> Let us turn now at last to the great question of fact, Does the absolute exist or not? . . . I may sum up that discussion by saying that whether there really be an absolute or not, no one makes himself absurd or self-contradictory by doubting or denying it.
>
> (James, 1996/1909, pp. 109–110)

Introduction

A kind of conversation often follows when I present or discuss CSR and the foregoing ideas. The first article that I published in this area was a preliminary call for the study of religious experience and meaning (Cresswell, 2014), and the very first email I received in reference to the article could well be described as: Well, do you believe in God or not? That is, I receive the same questions that seemed to have been put to James that I identified in the epigraph opening this chapter: "Does the absolute exist or not?" My supposition is that people want to know if the absolute exists in the sense of "does God exist?" It is a conversation about whether or not God is real and what the foregoing tells us about ontology and metaphysics. James was clear that we can deny the presupposition that God exists and not be absurd or self-contradictory. This claim is interesting since James defends religion in other places and is known for his defense of religious belief. He would appear to have the opinion that one can believe and/or not believe and neither position is definitively wrong. In this chapter, I will address this question while avoiding a definitive answer because such is not a fruitful one for religious believers. An unwillingness to commit to one position over another is, in fact, a stronger position for believers.

A corresponding and parallel conversation tends to emerge from the researchers engaging in CSR. I am sure that they are not concerned with the existence of God and prefer to remain kindly agnostic. Authors such as Justin Barrett, Jesse Bering, and Robert McCauley respectively include Christian believers, atheists, and mysterious agnostics. While the notion of an absolute God is not an issue for them, there is another absolute that researchers in CSR may find challenging. It's the quality of their knowledge and commitment to scientific research methods, which are taken to be cumulatively adding up to produce knowledge. No researchers would be so presumptuous as to say that they know everything. There is, however, a faith in science to be a vehicle to truth and this is the absolute that comes into play to offer resistance to the claims put forward above. This quasi-religious commitment to scientific method has been critiqued and even referred to as "methodolatry" in reference to worshiping method as an idol (Polkinghorne, 1983; Slife & Williams, 1995). The foregoing work – especially Chapter 5 – plays with such methodolatry and attempts to undermine it in order to develop a more robust CSR. This chapter is also about how an unwillingness to fall into methodolatry is, in fact, a stronger position for researchers.

James writes that "human institutions, by the purely technical and professional manner in which they come to be administered, end by becoming obstacles to the very purposes which their founders had in view" (1996/1909, p. 96). CSR seeks to understand religious belief, and I argued that it bypasses religious phenomena by being "purely technical" in its approach, becoming an obstacle to its own agenda. I tried to offer a way to expand CSR and so open up new avenues for research that do not suffer from such a bypass. As the saying goes, however, "the knife cuts both ways." Religious belief can take on a technical and professional manner that also bypasses dynamic and authentic religious belief. When religious belief becomes technical dogma of certain knowledge, it also bypasses something, and I argue that it is the dynamic quality of authentic belief that is bypassed. Where a religious believer may dogmatically adhere to supposed characteristics of God, scientific efforts may be hampered by methodolatry. This chapter is thereby about how letting go of both modes of certainty absolute is not a big concern and in fact, leads to a more robust science and/or expression of religious belief.

In what follows, I will review James' discussion of how we live in a pluralistic universe, which is a universe without certain appeal to the absolute and one of competing realities. This discussion raises the familiar challenges to CSR, but it also raises challenges to religious believers. Following the presentation of James' discussion of the pluralistic universe, I discuss how religious believers need not feel their belief in some sort of absolute is threatened. I will discuss how a certain belief in the absolute in religion cuts off understanding of faith just like it does in CSR.

Messing with certainty

Chapter 1 outlined how James positioned himself against Kant and Descartes, who had left us with a perspective on religious belief that treated it as a subjective property of mind. He contradicted them with the claim that human mental phenomena seem to come naturally as if by instinctual habit. With such terms, he did not mean that we have many hard-wired instincts that amount to evolutionarily shaped fixed action patterns (c.f. Friesen & Cresswell, 2017). Phenomena like belief amount to dispositional modus operandi where it feels natural to enact belief in a particular manner. Beliefs that seem natural are not abstract ethereal things locked inside the subjectivity of isolated monads. It is in this context that religious belief, like all mental faculties, is treated as an activity and not an entity (Chapters 3 and 4). It does not exist as an abstract thing because it is always drawn back to the world through its constitutive quality: the function of knowing *about* something (Chapter 2). One cannot consider "belief" per se because there is no belief without a sense of what the belief is about. That is, belief is always about something, and the implication is that it is always in relation to something. This feature of belief is very important because the things that our beliefs are naturally about always lead to the world and others (Chapters 2 and 4).

This quality of belief highlights how it is a relational phenomenon, and I discussed this at length in Chapters 1 and 2. One can have belief that includes knowledge about another thought or another thing, but it always involves a relation to another something. We do not have any element of belief that is isolated and on its own. That is, every religious belief is related to something else, and it simply does not make sense to talk about a belief that stands as a subjective isomorphic thing. James thereby provides substantial innovation for how we think of ideas like subjective mind and objective reality. He wrote:

> Just so, I maintain, does a given undivided portion of experience, taken in one context of associates, play the part of the knower, of a state of mind, of 'consciousness'; while in a different context the same undivided bit of experience plays the part of the thing known, of an 'objective' content . . . since it

> can figure in both groups simultaneously we have every right to speak of it as subjective and objective both at once.
>
> (1996/1909, p. 10)

A belief in another belief can be subjective, but the web of relations involved in belief never stops there. The upshot is that all religious belief eventually comes into relation with something in the world we share with others. Eventually, something that can be misconceived in terms of self-contained subjectivity is necessarily in relation to something that is not just inside of our heads. To paraphrase James, if one dies and stops believing in something, the object does not disappear, and it continues on without any single individual: ". . . I could perfectly well define [belief], what the knowing actually and practically amounts to – leading towards, namely, and terminating in percepts, through a series of transitional experiences which the world supplies" (1996/1909, p. 25). We can't say that religious belief is just subjective because of the way that it is a relational phenomenon: what it is depends on relationships to other things.

James also pointed out that belief is inherently experiential, by which he means that belief is experienced in the body by way of physiology and emotion (Chapters 3 and 4). While CSR promotes the idea that raw sensation is sifted into perception, James has long ago argued that the world is shaped and structured for us by virtue of our participation in life with others. In the course of living with others, our belief is formed in an embodied way. The entire way in which reality shows up is shaped in our beliefs formed through relations to others in our corporately entwined inter-embodiment. Religious belief is an experience of what we bodily feel as real in the world. The consequence is that it has a kind of naturalness that manifests in experience in terms of things like instinctual habits. It is natural in the sense of it feels like, as we say, "second-nature" (see Corcoran, 2009). We are talking about lived belief as a kind of instinctual or habitual knowledge of the world. Most of our life is lived in moments where we enact our beliefs through our bodies as a natural matter of course. Our knowledge is not an abstracted one because it is experiential. We know and shape a world experientially and not propositionally.

An implication of this approach is that any academic study is about the study of different worlds of experience. James (1996/1912) puts forward the notion of "radical empiricism" to help us realize that the facts of experience are entwined with a community that constitutes the world its members enact. One tries to grasp the realities of others, and this endeavor means that there are multiple worlds at play. There may be a religious world and the world of science as established by cognitive scientists, for example. Radical empiricism is to look to such realities and to understanding the lived worlds of others. A challenge for researchers is to recognize that there may be other worlds that do not share the same taken-for-granted realities. It is for this reason that James treats radical empiricism as the name given to the pluralistic rival of absolutism. Absolutism of any sort, says James, takes the approach that there is one singular world to understand, and absolutist research would be about identifying it.

Research wherein we take the presence of multiple worlds seriously has a different set of presuppositions insofar as "[e]very end, reason, motive, object of desire or aversion, ground of sorrow or joy that we feel is in the world of finite multifariousness, for only in that world does anything really happen, only there do events come to pass" (James, 1996/1909, p. 49). God, for example, is one part of a pluralistic system: "The God of our popular Christianity is but one member of a pluralistic system. He and we stand outside of each other, just as the devil, the saints, and the angels stand outside of both of us" (James, 1996/1909, p. 110). There are ambiguities inherent in the study of multiple worlds and radical empiricism involves recognizing that work often includes opposites and ambiguities. Hence, James writes that "[p]luralistic empiricism knows that everything is in an environment, a surrounding world of other things, and that if you leave it to work there it will inevitably meet with friction and opposition from its neighbors" (James, 1996/1909, p. 90). There are multiple ways of seeing the world, and this should lead to incongruences, which is crucial for good research (Chapter 5).

This vision for research leads to a mode that eschews certainty of the sort that methodolatry can pursue. Radical empiricism means that we have to think in fuzzier categories and in ambiguous spaces of uncertainty. James argues that the "vice" of humans is that they slide into a concern for seeking definitive answers and miss the "intermediate shades" that populate life. He writes that one

> cannot enter the phenomenal world with the notion of it in your grasp, and name beforehand any detail which you are likely to meet there. Whatever the details of experience may prove to be, *after the fact of them* the absolute will adopt them. It is an hypothesis that functions retrospectively only, not prospectively.
>
> (James, 1996/1909, p. 126)

Such an approach stands in contrast to research that seeks to close things down. It is for this reason that James suggests that radical empiricism stands for "the legitimacy of the notion of *some*" (ibid., p. 79) and that we can know something about our subject matter. That is, we do not have definitive or complete knowledge:

> radical empiricism allows that the absolute sum-total of things may never be actually experienced or realized in that shape at all, and that a disseminated, distributed, or incompletely unified appearance is the only form that reality may yet have achieved.
>
> (ibid., p. 44)

This approach opens up a helpful discussion in relation to scientists who are ostensibly concerned with questions of ontology. The foregoing discussion does not compel a researcher or believer to take a firm ontological stance on the topic of the real existence of God. Such apposition would also seem to undermine religious believers, but the next section will discuss how faithfulness to a religious practice in everyday life is far less bound up in ontological certainties.

Danger equals certainty in science and/or religion

Giovanni Vattimo (2002, 2011) elaborates such notions by arguing that uncertainty is important because certainty about ontological and metaphysical claims leads to violence. This claim is grounded in how such ontological claims are rooted in what he calls a "truth-totality," which is a sure and certain knowledge about whatever subject matter is at stake. One who takes a position of truth-totality is one who takes his or her knowledge to be complete and certain about a given domain. This kind of modus operandi would be marked by a religious practitioner who claims to know everything about the character and quality of God or appeals to universal systems of values that spell out orthodox behavior. Vattimo notes how absolutist systems close down dialogue because they include one-way didactic knowledge where some people have privilege over others. He also notes that "there remains the problem, specifically democratic, of keeping the unchecked power of the technicians in the various areas of social life from simply replacing the authority of the philosopher-kings" (p. 17). Technicians are those who claim expertise in a domain of practice wherein others have no voice and are subjugated. It is a mode where a privileged person's knowledge dominates over others and understanding is reduced to a technological mode. One who sets the formula dominates by way of compelling adherence to the formula such that the technological mode can also exist in a religious domain where it and actions in relation to it are a means to an end. It involves treating nature like something standing in reserve for us to use because we have no ethical obligation to it. This mode leaves others and religious texts as objects that serve whatever kind of ends we want. A technological relation to religion involves treating it as a standing reserve such that it is something we can plunder and strip without a sense of offending something that has an opinion. It is in this way that religious expression can take the same hegemonic and totalizing form as is problematic in CSR.

Whether religious or scientific, a problem with truth-totality is that it can impede human freedom, and therein lies violence. Vattimo writes that the

> attainment of instrumental rationality in mass society may realize totality, but it does so in a way that is the opposite of liberatory. . . . totality, which appears to be the sole possibility of access to truth, is the polar opposite of the liberty that ought to accompany truth.
>
> (2011, p. xxvii)

When one becomes an expert, one shuts down the potential for authentic dialogue and dehumanizes the other. To cut off one from a place to voice one's experience is to perpetuate a kind of violence that was discussed in Chapter 5 in terms of epistemic violence. In another place, he argues that such technological modes involve identification of historically situated laws and procedures as being nature, as in an irreducible feature of reality (2002). Unquestionable dogmatism resonates with a technological mode and Vattimo writes that "[t]he hidden that tends to vanish in the specialization of the sciences is Being qua Being, the integrality of individual and social experience that needs to be protected from technological schizophrenia and the lapse into authoritarianism that it entails" (2011, p. 21). The kind of person one is becomes dictated to one by scientists and/or religious believers as what one sees, and experience is undermined in favor of the laws specified by truth-totality. In reference to the Christian tradition, he argues that violence rules as long as Christianity is linked to truth-totalities (2002, p. 116). Hence, we cannot lose sight of the way that certain religious or scientific perspectives run the risk of committing violence.

Vattimo (2002, 2011) is famous for his discussion of the "weakening" of truth, which amounts to the demise of appeal to uncertain truth claims that a technological mode perpetuates. Even in science, we have had to come to grips with how the rise and fall of science paradigms are complex historical events. That is, "[t]ruth faces a challenge in the world of postmodern pluralism – the challenge of coming to grips with the fact that consensus on individual questions is above all a problem of collective interpretation, of constructing paradigms shared or, at any rate, explicitly recognized" (2011, p. xxxv). Much like I argued in Chapter 2, he raises the issue that truth claims are constituted in communities in a deeply pervasive way that feels personal. That is "truth is not encountered but constructed with consensus and respect for the liberty of everyone, and the diverse communities that live together, without blending, in a free society" (2011, p. xxxvi). In reference to religious belief, I outlined how cultural and linguistic practices resulted in the constitution of experienced realities. Religious believers cannot deny the inherent pluralism in a global society. The consequences of denying such pluralism can be dire because continuing to "think of faith as a deposit of truths more certain than those ascertained by science, continues to profess a more or less literal creationism, and continues to try to impose its own biblical anthropology on the State, it is fated to dwindle away, in a world where science and human rights are becoming universal" (2011, p. 53). An allegiance to unassailable religious metaphysics is untenable just as is allegiance to scientific methodolatry.

This demise of truth-totality is a good thing, argues Vattimo (2002, 2011), because it enables weakening of truth, which in turn enables a fuller expression of religion. He writes the following about the Christian tradition: "nihilism is the postmodern interpretation, or version, of Christianity – in my opinion the only one that can save it from ebbing away to nothing or ending violently in a universal religious conflict" (2011, p. 58). Secular pluralism is a phenomenon Vattimo sees as rooted in Nietzsche, who wrote in the *Gay Science* that "God is dead" and it is worth quoting this section at length:

> Have you not heard of that madman who lit a lantern in the bright morning hours, ran to the market place, and cried incessantly: "I seek God! I seek God!" – As many of those who did not believe in God were standing around just then, he provoked much laughter. Has he got lost? asked one. Did he lose his way like a child? asked another. Or is he hiding? Is he afraid of us? Has he gone on a voyage? emigrated? – Thus they yelled and laughed. The madman

> jumped into their midst and pierced them with his eyes. "Whither is God?" he cried; "I will tell you. *We have killed him* – you and I. All of us are his murderers. But how did we do this? How could we drink up the sea? Who gave us the sponge to wipe away the entire horizon? What were we doing when we unchained this earth from its sun? Whither is it moving now? Whither are we moving? Away from all suns? Are we not plunging continually? Backward, sideward, forward, in all directions? Is there still any up or down? Are we not straying, as through an infinite nothing? Do we not feel the breath of empty space? Has it not become colder? Is not night continually closing in on us? Do we not need to light lanterns in the morning? Do we hear nothing as yet of the noise of the gravediggers who are burying God? Do we smell nothing as yet of the divine decomposition? Gods, too, decompose. God is dead. God remains dead. And we have killed him. How shall we comfort ourselves, the murderers of all murderers? What was holiest and mightiest of all that the world has yet owned has bled to death under our knives: who will wipe this blood off us? What water is there for us to clean ourselves? What festivals of atonement, what sacred games shall we have to invent? Is not the greatness of this deed too great for us? Must we ourselves not become gods simply to appear worthy of it? There has never been a greater deed; and whoever is born after us – for the sake of this deed he will belong to a higher history than all history hitherto." Here the madman fell silent and looked again at his listeners; and they, too, were silent and stared at him in astonishment. At last he threw his lantern on the ground, and it broke into pieces and went out. "I have come too early," he said then; "my time is not yet. This tremendous event is still on its way, still wandering; it has not yet reached the ears of men. Lightning and thunder require time; the light of the stars requires time; deeds, though done, still require time to be seen and heard. This deed is still more distant from them than most distant stars – *and yet they have done it themselves*.
>
> (Nietzsche, 1964, pp. 94–95)

It may be tempting to consider this text to entail an atheist claim, but Vattimo (2002) argues that this is not a good interpretation. It is a mode of being marked by certainty that kills god because one lives a life related to the dogma and not to a dynamic active being. It is in this sense that the death of God is the only way to avoid violence. That is, the speech means that we bracket out the presence of the divine when we turn to a technological mode of being and rules. It is in this way that secularism, argues Vattimo (2002) involves an end of unassailable truth-totalities that paradoxically enables a return to the sacred – a return to a dynamic lived religious expression instead of (religious or not) technological modes.

A challenge, then, is how to approach truth if it has been weakened in light of a pluralistic universe. A pluralistic world outlined by James and Vattimo could seem to put us in a place where we collapse into something like an "anything goes" situation but we are not, however, in a situation where "anything goes" on two bases.

One is rethinking of relativism along the lines of the Christian idea of kenosis, or humbling of oneself. Vattimo (2011) highlights that the notion of kenosis in the Christian tradition actually gives space to pluralism. One can see how the notion echoes the theme of pluralism when Vattimo writes that to

> speak of a "kenotic" or relativistic God is to fully accept that the age of the Bible as a deposit of knowledge, the truth of which is guaranteed by divine authority, is over and gone, and that this is not an evil to which Christianity must temporarily adapt while waiting for the moment of revanche but is part of salvation history.
>
> (2011, p. 53)

The notion comes from the act of Christ where he humbled (kenosis) himself to be a human and die on the cross. The notion is not about an historical claim because it is about a kind of action that humans can imitate. It is about surrendering one's own perspective enough to be like another and understand another. Kenosis entails suspending the temptation to act out of truth-totality and to remove blinders that can hinder people from understanding others. It is about a genuine love and emptying oneself that is not a self-annihilation because annihilation of the self is false humility. While this is a Christian notion, one would be hard pressed to find a religious tradition that would eschew it. Moreover, the foregoing argument in reference to CSR amounts to a kenotic approach to research.

This approach to religion radically changes how we think of relativism because it does not primarily undermine truth. It is clear that a conception of something divine

> can no longer be the God of definitive and absolute truth that allows no doctrinal variation. That is why he may be called a "relativistic" God. A "weak" God, if you will, who does not show us our weakness as a way of asserting (against rational expectation, as a mystery to which we must submit, as the ecclesiastical discipline we must accept) that he is luminous, omnipotent, sovereign, and awesome, that he bears all the traits of the threatening and reassuring personage of natural-metaphysical religiosity.
>
> (2011, p. 55)

The kenotic side to this is that Nietzsche's death of God points to an event that transforms existence of the person who becomes humble and open to apprehending the other. An end to certainty "is an event that cannot be ascertained objectively, one to which thought is called to respond. It is an event that transforms the existence of the person who receives the announcement . . ." (Vattimo, 2002, p. 13). The kenotic approach rests upon the weakening of God that paves the way to a fuller religious life.

A divine wholly other is possible in light of weakening because authentic relationality is enabled as dialogue is enabled. Truth-totality is weakened so that we have a voice in reference to the divine:

> At this point it becomes impossible to think Being as objective rationality because, thought in that way, it would just be the ground of the dehumanization of the world, where all that exists is functionality predetermined by a colossal mindless mechanism. Seen in this light, the kenosis that is the original meaning of Christianity signifies that salvation lies above all in breaking the identification of God with the order of the real world, in distinguishing God from (metaphysical) Being understood as objectivity, necessary rationality, foundation.
>
> (2011, p. 55)

Weakening of truth is parallel to James' pluralism and leads us to think differently about research and religious experience. Religious belief can only have meaning when characterized by deep uncertainty of opinion because it is only then that there is any kind of liberty and equal dialogue.

Another basis preventing us from falling into an "anything goes" situation is that the weakening of truth does not necessitate a kind of relativism demarcated by lassé-faire relativism. Just because what people think that they know and experience about God is not absolutely certain, for example, does not mean that they must stop believing in God with conviction (Vattimo, 2002, p. 6). In fact, the pluralistic worlds introduced by James and the weakening of truth permit us to come back to the importance of the present and how life is marked by an "event of Being." To understand what this event is about, it is important to first be clear that we are dealing with a present orientation to life and how we are a community of people who constitute what is taken to be real (Chapter 2). Vattimo writes that " 'ontology' . . . denotes the thought of Being in both senses, subjective and objective, of the genitive. This decision alone on the meaning of the term marks a profound difference [from] most ontologists, who reduce ontology to a theory of objects" (2011, p. 23). To say that ontology includes objectivity – scientific or religious – is not to say that we are dealing with a class of things independent of us that must be discovered as in the case of a "theory of objects." It includes the objects that people jointly constitute as real in a deeply experiential way that I outlined in previous chapters. It means that our experience is not separable from things – and so includes subjectivity – but it is not reducible to a single individual. Thereby,

> it becomes impossible to think Being as objective rationality because, thought in that way, it would just be the ground of the dehumanization of the world, where all that exists is functionality predetermined by a colossal mindless mechanism. Seen in this light, the kenosis that is the original meaning of Christianity signifies that salvation lies above all in breaking the identification of God with the order of the real world, in distinguishing God from (metaphysical) Being understood as objectivity, necessary rationality, foundation.
>
> (2011, p. 55)

We are not dealing with worlds that can be played with as if they are a free array of options. The world has experiential gravitas in its verisimilar reality. A kenotic approach recognizes such deep experiential meaning and treats the world of the others as something as serious as one's own.

As such, when we start to consider the truth of our work, we do so in the context of understanding an other. Vattimo writes that "[w]hat we might call the epistemological precondition of social and intercultural dialogue is precisely this truth of horizon, which politics has the task of grasping and attempting to make explicit and to construct" (2011, p. xxxiii). We come into contact with others and so come into contact with other worlds as James describes them. The weakening of truth means that, instead of managing and conveying a truth of one's own perspective that is taken as unquestioningly universal, one reflects on the socio-historical locatedness of one's world of experience:

> Light can be shed on the problem, it seems to me, if we focus on the twofold significance of the ontology of actuality: making oneself aware of the paradigm into which one has been thrown yet suspending its claim to definitive validity and heeding Being as that which remains unsaid.
>
> (ibid., p. 32).

It becomes a potential object of questioning that opens up the potentiality of understanding another's world of experience. Such experience is the mark of late modernity where we embrace the end of metaphysics to really understand what it means to be an other (see Vattimo, 2002, p. 34). It is for this reason that we cannot get at an essence outside of experience, but the relationship between I and an other provides a purview into a different lived essence.

Such activity is not the easiest, and so this discussion leads to one of interpretation wherein scientists and/or religious believers must learn how to interpret an other. An important issue is that all interpretation is not a matter of distancing oneself from the world of experience one lives, which methodolatry requires. That is, "interpretation is the idea that knowledge is not the pure mirroring of the datum but an interested approach to the world using schemes that have varied over time" (Vattimo, 2011, p. 66). James' discussion of plural worlds makes this clear and it is only reinforced by Vattimo's discussion of how religious life is deeply experiential. If we could simply surrender our worlds of experience, we would then fall back into the "anything goes" relativism instead of the kenotic kind I propose above. Interpretation involves a phenomenological analysis of religious experiences of oneself and an other. It means to engage an other by imagining his world of experience as "the world and someone who stands in the world and takes his bearings in it utilizing his cognoscitive capacities, in other words choosing, reorganizing, substituting, and so on" (ibid., p. 63). Since, as I articulated in the foregoing, experience is bound up in communal participation, we are looking not just at the experience of a single other when we engage in the interpretation of his experience. We are engaging in him as an expression of whatever communities he lives and so

> the question of truth is recognized as a question of interpretation, of the application of paradigms that, in turn, are not objective (since no one verifies or falsifies them except on the basis of other paradigms) but that are a matter of social sharing.
>
> (ibid., p. xxxiv)

The reality of what we know together is therein an issue of interpreting each other as best we can.

The best part of this endeavor is what interpreting amounts to. Authors such as Vattimo, and others (e.g., Bakhtin, 1990; Taylor, 1989) note that we often live our religious or scientific worlds of experience as second nature. Our own taken-for-granted understandings sink into obscure tacit behavior because we naturally act without a second thought. We cannot see ourselves and the background presuppositions in which we ground action. Interpretation leads to a situation of relative juxtaposition of worlds of experience. When we understand an other, we see him from a perspective that is outside that of his own, and we simultaneously come to see him from within his reality. We conversely can see our own taken-for-granted realities in the same process. It is for this reason that Vattimo writes about interpretation as the "opening of a horizon (or paradigm) that enables the truth of statements that conform to things, the truth of propositions verified or falsified" (2011, p. xxx). It is against the exposed realities that people naturally live that the truth of a claim can be verified or tested. Rather than testing truths against truth-totalizing facts, we test them against different realities that are uncovered in interpretation. Interpretation and a weak approach to truth sets us free from what we would normally do as a matter of course. Hence, Vattimo writes that the "truth that sets us free is true precisely because it sets us free: if it doesn't free us, it is disposable. That is why I refuse to admit that weak thought, with all that it entails, is just a species of preaching of tolerance" (ibid., p. 79). To really experience freedom is to have the realities that constrain us challenged. The situation is not mere tolerance or "anything goes" relativism because worlds of experience are at stake, which they should be if we are to avoid epistemic violence. Such a point brings more gravitas to the notion that interpretation does not mean that we are dealing with a kind of relativism where "anything goes":

> So it isn't true that if we accept the prevalence of interpretation, anything goes and anyone can say whatever they please. The rules don't vanish, but they are the rules that apply in that language . . . if you are going to preach a religion of free creation, salvation, grace, and miracles, then objectivity goes out the window, because definitive, metaphysical objectivity is the structure that, once discovered, we know (and cannot not know) to be the way it is (and cannot not be).
>
> (Vattimo, 2011, p. 68)

Interpretation involves a turn to moral contemplation and inner conversation as was outlined in Chapter 5 (see also Vattimo, 2002, p. 44). The stakes are high as we bring worlds into contact and lay bare what we take to be the very fabric of reality.

A principle that guides such endeavors is thereby not a search for propositions conformable to facts, be they religious or scientific. It is the principle of charity, which one finds in many religious traditions (James, 1982/1902). Charity is no longer dependent, however, on truth-totality as a rule for action. Taking the risk of being entailed in weak truth and engaging in interpretation involves a lived charity where one seeks to understand another. It requires a radical suspension of one's own purviews as one realizes one's weak quality. That is,

> The last thing you can be certain of is your most deeply rooted certainties, because those are the ones you were taught by your aunts and grandmothers, the Church, the authorities, the newspapers, advertising. Christianity is marching in a direction that can only be that of lightening and weakening its burden of dogma in favor of its practical and moral teaching. In that sense too, charity takes the place of truth.
>
> (Vattimo, 2011, p. 78)

Hence, the death of God is what enables a consideration of heritage and communities. Truth becomes relational phenomena most acutely found in the development of friendship:

> But the golden rule [i.e. do to others what you would have them do to you] operative here necessitates an intentional attentive love. If I totally give myself up to the text I am reading or lazily allow its words to roll over me, I fail ethically in that I bring nothing of myself to the encounter.
>
> (Contino & Felch, 2000, p. 6)

Interpretation must follow charitable imperative, and the charitable trait is attentiveness. This rich form of charity is enabled precisely because "we are living in an epoch that through science and technology can dispense with metaphysics and a metaphysical God: a nihilist epoch. An epoch in which our religiosity can finally develop into the form of charity no longer dependent on truth" (Vattimo, 2011, p. 59). Instead of an objecting moral order, we turn to care for an other, and this turn amounts to a dynamic activity negotiating friendship and struggling dispositional hostility to a world different from our own. (Vattimo, 2002, p. 105). This central importance of the other means that charity becomes the central factor of truth if we leave the objective view (ibid., p. 111).

Concluding

James' approach of plural worlds may open up a challenge to those in CSR because it challenges their scientific approach. It exposes it as one worldview among many. This move is not so bad as we move forward to a more dynamic CSR that can account for people's experience. The relativism exposed by a pragmatic cultural psychology of religion also opens up concerns for believers. Far from being a liability, it enables an interpretive approach grounded in charity.

Conclusion

Abstract

This concluding chapter expands on James' approach to truth to offer a vision for the roles of researchers in academic work. I argue that theory cannot be avoided by researchers because being "simply" pragmatic can be disingenuous, which is why the foregoing was an attempt and careful discussion of theory in the cognitive science of religion. Discussions about theory are crucially important because they challenge simplistic or reductionist answers, and this is a skill that academic researchers need as the cloistered intellectual is becoming untenable. Being open to theory while practicing the skill of understanding the life worlds of our participants does not mean uncritical acceptance of whatever research participants say. I cast the vision for an academic researcher to be an ambivalent figure – a figure in exile yet not wholly disconnected. A researcher in the cognitive science of religion can be one who moves across barriers and so is a specialist in making connections by being both in and outside of the communities studied.

Recap

I started with the claim that William James shows us how religious belief works in a way that is concrete and embodied. This approach raises the challenge that the truth claims in CSR are about general naturalistic mechanisms that don't bear much relevance to religious beliefs as they show up in life. Rather than presuming natural cognitive mechanisms to be the only feature of belief that matters, I have attempted to show that we can return to James because he offers a direction that expands the notion of scientific truth. One way to approach scientific truth is to consider how a claim fits with life, which is about how truth is inseparably bound to life because people do not use truth in an abstract and disembodied sense. This means that truth is not used in a vacuum because it is used in relation to all of the other aspects of life. Importantly, a scientific claim must resonate with the practices of a community and so is tied to life outside of the mind. Religious belief emerges in the interconnected milieu of life, and so a claim about such belief must be part of this milieu to be true. Truth is not just separate from life in the abstract and disembodied sense and a belief is true because of how it relates to emotional valuations, ideas, others, and things that people are engaged with. James' view of truth was that a scientific claim is true when it fits into the flow of life insofar as it fits into the complex web of interrelations constituting experience. A claim made by a scientist in CSR gains the status of truth beyond the reality of scientists the moment that it falls into resonance with the reality of those she is proclaiming a truth to.

Chapter 2 elaborated on this idea by arguing how culturally oriented research can get at religious belief in a way that fits with the world of religious believers. Culture is necessary for a discussion of belief if one is to find a fit with culturally shaped life: to truly understand belief is to grasp the culturally shaped stories that are tied to it. Taking culture, language, and narratives seriously enables researchers to get at realities lived by others because it is through these serious aspects of life that such realities are constituted. CSR has yet to develop an account of culturally shaped religious belief because of its naturalist approach, but I hope that the foregoing can help bridge this gap.

Embedded in this discussion of culture is a second criterion for truth: fit with a community. As I mentioned, it looks like objectivity is lost if we worry less about universal cognitive mechanisms, and this perception is partly true, but not disastrous. Facile thinking in research leads scholars to conclude that a collapse into subjective fancy does not necessarily follow. I showed how truth is not a matter of the mere proclivities of subjectivity when one takes a cultural approach. Standards of a truth claim's fit are socio-normative in quality and are integrally integrated with life with others. One cannot have a true belief on one's own. Beliefs have truth by virtue of fit and generativity within communities. A true belief is one that fits with the standards of a community, and because there are different communities in the world, there are different standards of truth.

Chapter 3 took up a discussion of this criterion to offer a place to start for researchers in CSR. I presented how child development involves learning to embody communal standards and how the cultivation of cognition in infancy can shape religious belief and the meaningful experience it entails. Since such cultivation of cognition involves emotional signaling and the importance of interrelationships, I offered an approach to cognition that includes socially shaped religious belief. The chapter thereby outlined changes necessary for researchers to broaden our science by providing an approach to cognition.

A third criterion for scientific truth is generativity insofar as, for a belief to remain true, it must generate something for us as we move forward such that it makes future activities smoother. Human life is highly dynamic, and so enshrined dogma about cognition or any other domain bypasses such dynamics. We change and the world around us changes such that a true belief could be antiquated and irrelevant. A true belief helps us succeed in life as opposed to creating friction and tension. I outlined how CSR researchers have been moving forward together insofar as their work has generated lots of ideas and discoveries among themselves. A challenge I attempted to address is that such work has not been generative for religious believers due to abstract and disembodied naturalism. The goal of Chapter 4 was to broaden the way we approach the study of religious belief so that we can move forward in a broader range of circles.

This discussion raised a challenge, which is how to create a generative interface between a strong cultural approach outlined in Chapter 2 and an approach to cognition. Chapter 4 further explored a unique approach to the evolution of human cognition with the aim of providing an approach that better includes the cultural quality of religious belief outlined in Chapter 2 and articulated in Chapter 3. An enactive approach can aid in this exploration because it offers a phenomenologically oriented view of the evolution of human cognition that is *fundamentally interdependent* with the surrounding socio-cultural milieu. Instead of treating genes as the evolutionary unit of transmission, I offered a vision of cognition that sees the unit of evolution as the selection for and passing on of dynamic cultural belief systems. This work thereby expands our understanding of human evolution by highlighting how religious belief emerges in interdependent dynamic cultural systems. This is an approach to cognition that is not abstract and disembodied while being generative via its theory of cognition that can account for religious belief. As such, I attempted to build a generative bridge between CSR and religious believers.

Chapters 5 and 6 represented a shift to deal with the kinds of implications that flowed from the foregoing. I outlined a vision for how psychologists ought to go about their research of cognition by discussing the notion of virtue. That is, a CSR that does not alienate itself from religious believers should include standards of good research, and so I sought to grapple with this issue. I did so by expanding what we take good research to be through a discussion of virtues needed to be a good researcher. This discussion flows into the one in Chapter 6 where I grappled with the tensions that come from rejecting universal transcended truths. A culturally informed CSR that speaks to both social scientists and religious practitioners is possible. The indeterminacy introduced by the foregoing is not threatening to research on religious experience or researchers who work in the CSR. In fact, it is an inoculation against a kind of certainty that comes from appeal to metaphysics – whether they be religious or scientific. The deconstruction of authoritative experience that is part of the proposed approach makes room for dynamic personal experiences and a vibrant self-correcting science.

Role of academic researchers

One of the struggles that I face with colleagues is a reticence to get deeply involved with theory. I often see theory take a backseat to practice because theory is tied to an image of an esoteric professor working in a dusty office using concepts irrelevant to the people in the real world. Such reticence is a little seditious insofar as it is a position that theory has to be tested and this is really a denial of how theory sets the conditions for what counts as observable evidence (see Chirkov, 2016; Gergen, 1985). Both are rooted in an effort to value practical life and working class critique of bourgeoisie culture – elitist notions of being cultured in terms of art and philosophy (see Shor, 1987). Theory is a casualty of such modes of thought that leads to an erroneous split between theory and practice. Many public intellectuals have noted, much like Žižec (2008), that theory permeates everything, down to the ways that we design our toilets. Theory is always at play, and so avoiding it to be "simply" pragmatic is disingenuous or self-deceiving at best or oppressive at worst.

My efforts have been to leverage a trend among theoretical psychologists and historians that notes how those in the human and social sciences have made themselves amenable to the same thinking (e.g., Brown & Stenner, 2009; Danziger, 1994; Stam, 2015a; Teo, 2008). I started with pragmatism to provoke a deep conversation about the role of theory in CSR. CSR has pragmatic value to researchers in the area because they enact a generic form of theory. They remain largely mute outside the academy who shares these presuppositions, and so I attempted to raise a discussion that I don't often see in the CSR. That is, debates about the phenomena of study are often not presented but they are exceptionally important. We can generate results of studies but our theory runs the risk of setting up the parameters for what these results mean. Cromby (2015) describes how we can dangerously confuse the apparatus of our research with the results. The theory that sets the experiments and methods often constitute the apparatus (i.e., the language game played by researchers). We can then think we have "found" some sort of results when they are epiphenomena of our theory. My case has been that researchers in CSR need to examine theory in order to have wider significance beyond its own community.

What might seem like a paradox is that my agenda is rooted in a resistance to bourgeoisie academia that undergirds resistance to theory. I have attempted to push this pragmatism to a reasonable depth by taking up the cause that academic work ought to be authentically engaged in community. The removed and cloistered academia is swiftly becoming untenable as there are increased efforts directed at vocational training (see Nussbaum, 2016). Rather than promoting vocational training as the answer to this problem, I am suggesting a pragmatic approach that retains value of the liberal arts and how academic endeavors are about a vital role in society as opposed to only vocational training.

This role has been described by Nussbaum (2016) in terms of how academic efforts are about creating independent thinkers in contrast to "producing generations of useful machines, rather than complete citizens who can think for themselves, criticize tradition, and understand the significance of another person's sufferings and achievements" (p. 2). The kind of theoretical engagement that I'm promoting attempts to lead to the

> ability to think well about a wide range of cultures, groups, and nations in context of a grasp of the global economy and of the history of many national and group interactions is crucial in order to enable democracies to deal responsibly with the problems we currently face as members of an interdependent world. And the ability to imagine the experience of another – a capacity almost all human beings possess in some form – needs to be greatly enhanced and refined if we are to have any hope of sustaining decent institutions across the many divisions that any modern society contains.
>
> (p. 10)

The kind of theoretical engagement I propose is the kind that is about questioning and challenging simplistic reductionist answers.

The kind of CSR that I am attempting to provoke is one that is not from an alienating outsider position that is pragmatically mute. Following Nussbaum (2016), this is a call to engage people who are fully real persons and not objects (i.e., subjects of research in the sense of persons subjectified to research methods) in science experiments. This role is to engage in work that include reflexivity and awareness of positional thinking (see pp. 36–37). A result is a milieu of accountability where differently positioned people are engaged in interactions and account to others who may not share theory. Nussbaum inspires the necessity of liberal arts education, and this book has been about how such a call extends to research.

Researchers should not, however, take a position of uncritical acceptance of whatever research participants say, and Edward Said (1996) outlines why this is the case. He asserts that an academic researcher is something of an outsider in society and uses the notion of exile to express this idea:

> The pattern that sets the course for the intellectual as outsider is best exemplified by the condition of exile, the state of never being fully adjusted, always feeling outside the chatty, familiar work inhabited by natives, so to speak, tending to avoid and even dislike the trappings of accommodation and national well-being.
>
> (p. 53)

An exile is not quite in step with society, and so an implication is that a researcher could be conceived as someone who does not always see the world in the same way as many around him – always questioning what others seemingly take for granted. A question that this conceptualization of an intellectual begs is: From whence does this exile come about?

Said is not entirely clear where exile comes from, but I see Bakhtin (1981) offering some insight. Exile comes about from the very nature of an intellectual vocation that exposes one to many voices and perspectives. I would describe this as the polyphonic condition of intellectual life because polyphony expresses a condition of being able to speak from within multiple languages (see Cresswell, 2011). Languages, in this case, involve genres of communication that constitute the ways participants in a community see a world. Within any language, there is a way of seeing the world that shapes the ontologies participants in a language experience. Academic life ought to be polyphonic as far as it involves exposure to many languages by virtue of one's engagement with others in conferences and partnerships and engagement with theorists both past and present through the discipline of reading challenging academic print media. One who ceases to rigorously explore new languages through broad and challenging theory is not an intellectual (this does not mean that one who does not read such does not belong at a university; it just means that such a one cannot consider the moniker of intellectual appropriate). Academic life amounts to rigorous and courageous explorations of different languages and thereby different ontologies. This vision is consistent with Said, who writes that an intellectual life involves "... the pleasure of being surprised, of never taking for granted, of learning to make do in circumstances of shaky instability that would confound and terrify most people. An intellectual is fundamentally about knowledge and freedom" (1996, p. 59). This exploration of languages is what makes an intellectual an exile because the freedom of knowledge ruptures a potential of falling in step with the zeitgeist of one's time.

An important feature of intellectual exile foreshadowed above is an intellectual's capacity to pursue freedom in the form of challenging power. A well-known concern is that power works in the daily embodied practices of life (e.g., Bakhtin, 1981; Bhabha, 1994). Languages constituting power are embodied without question and sink from vision in their ostensive reality. The polyphony that puts an intellectual into exile relativizes such languages. Where power is unquestioned within one language, it becomes questionable when different linguistically constituted ontologies come together. An intellectual's vocation enables him to see the world in ways of other than the status quo:

> Even if one is not an actual immigrant or expatriate, it is still possible to think as one, to imagine and investigate in spite of barriers, and always to move away from centralizing authorities towards the margin, where you see things that are usually lost on minds that have never traveled beyond the conventional and comfortable.
>
> (Said, 1996, p. 63)

An academic can see alternatives to an institutional habitus of power, and this is the potential I see for researchers in CSR. Thinking deeply about theory and exploring pragmatic modes of inquiry enables the quality of being an exile, which then enables one to respond to power. A CSR researcher can be understood as "set apart, someone able to speak the truth to power, a crusty eloquent, fantastically courageous and angry individual for whom no worldly power is too big and imposing to be criticised and pointedly taken to task" (ibid., p. 8).

A value of the vision of an academic articulated by Said (1996) and Nussbaum (2016) is that this vision sees an intellectual as an ambivalent figure – a figure in exile yet not wholly disconnected. A polyphonic experience leads an academic to be one that naturally crosses barriers and goes as far as to abhor their seeming in-traversable qualities. Consider how a researcher in CSR could be marked by a

> desire to be moved not by profit or reward but by love for an unquenchable interest in the larger picture, in making connections across lines and barriers, in refusing to be tied down to a speciality, in caring for ideas and values despite the restriction of a profession.
>
> (Said, 1996, p. 76)

A researcher in CSR can be one who moves across barriers in her play with languages and so is a specialist in making connections by being both in and outside of languages simultaneously. While a researcher crosses barriers by becoming immersed in languages, this movement includes populist communities in which an intellectual is situated. As such, an "intellectual's voice is lonely, but it has resonance only because it associates itself freely with the reality of a movement, the aspirations of a people, the common pursuit of a shared ideal" (ibid., p. 102). It is such participation in languages of surrounding populace communities, an intellectual finds resonance while remaining an exile that embodies a potentiality of emancipation:

> The intellectual's representations – what he or she represents and how those ideas are represented to an audience – are always tied to and ought to remain an organic part of an ongoing experience in society: *of the poor, the disadvantaged, the unrepresented, the powerless.*
>
> (ibid., p. 113; emphasis added)

A researcher can speak to power from his ambivalent tension as part of society yet exile to the status quo of hegemonic machinations. That being said, "[s]peaking truth to power is not Panglossian idealism: it is carefully weighing the alternatives, picking the right one, and then intelligently representing it where it can do the most good and cause the right change" (ibid., p. 102). It is precisely the exiled nature of a researcher's voice that challenges an oppressive status quo and precisely its identification with local community languages that enables the challenge of power from within – translating and bending languages to enable emancipation within and from such languages. *A researcher in CSR need not be removed from society and can engage its populace languages with understanding that still questions power.*

It is by way of polyphony that an academic researcher can question power. Weighing alternatives and voices includes bringing people into contact with ideas that they never would see otherwise. A researcher in CSR can bring people into contact with other ideas and so would be

> an individual with a specific public role in society that cannot be reduced simply to being a faceless professional, a competent member of a class just going about her/his business. The central fact . . . [is] that the intellectual is an individual endowed with a faculty for representing, embodying, articulating a message, a view, an attitude, a philosophy or opinion to, as well as for, a public. And this role has an edge to it, and cannot be played without a sense of being someone whose place is publicly to raise embarrassing questions, to confront orthodoxy and dogma (rather than produce them), to be someone who cannot not easily be co-opted by governments or corporations, and whose *raison d'être* is to represent all those people and issues that are routinely forgotten or swept under the rug.
>
> (ibid., p. 11)

It is this vision of the academic researcher challenging authority in all forms while not engaging in "opposition for opposition's sake. But it does mean asking questions, making distinctions, restoring to memory all those things that tend to be overlooked or walked past in the rush of collective judgment or action" (ibid., p. 33). A researcher in CSR advances both freedom – through engaging communities and asking questions of power – and knowledge – through the polyphonic background that opens oneself and others to new languages.

All of this is to say that the foregoing was an attempt to address theory in a way that is deeply pragmatic instead of a shallow pragmatism that avoids theory. James (1982/1902) offers an excellent place to start because he takes up the same task and so continues to have relevance today (see Cresswell et al., 2017). His psychological study of religious experience creates a rich approach to the phenomena. On the one hand, he sought to avoid discussion of the institution of religion, which can be an organizational study devoid of experiential gravitas. Making his discussion about experience highlights how religious institutions are founded on experience. On the other hand, James discussed "medical materialism" to raise that the problem of presupposing experience is easily reduced to biology. He pointed out that knowing the organic antecedents does not help us understand why one state of mind is any better than any other. All biological reductionism gives us is a destructive kind of dogmatism that prohibits us from understanding the phenomena. Knowing the origin of experience does not help us understand it from the purview of those living it, and so deep thinking about religion involves looking at the pragmatic value irrespective of origins. James challenges us to think deeply about religion with a view to its fruits, and my work has been to show how to do this.

It is important to think deeply about religious experience, James (1982/1902) asserts, because of how real it is. Attitudes are due to the objects of our consciousness, but such phenomena have very real and compelling force:

> Such ideas, and others equally abstract, form the background for all our facts, the fountain head of all the possibilities we conceive of. They give its 'nature,' as we call it, to every special thing. Everything we know is 'what' it is by sharing in the nature of one of these abstractions. We can never look directly at them for they are bodiless and featureless and footless, but we grasp all other things by their means, and in handling the real world we should be stricken with helplessness in just so forth as we might lose these mental objects, these adjectives and adverbs and predicates and heads of classification and conception.
>
> (p. 56)

What is taken as seriously real are these unseen compelling forces that impact life in serious ways. Whatever rational set of facts is put forward is grounded in unseen realities that need to be part of our CSR. James still speaks to us today by showing us how we are currently dealing with such realities in the study of religious experience, and researchers thereby need a vocabulary for them.

References

Anderson, J. (1998). Production systems and the ACT-R theory. In P. Thagard (Ed.), *Mind readings: Introductory selections on cognitive science* (pp. 59–76). Cambridge, MA: MIT Press.

Argyle, M. (2000). *Psychology and religion: An introduction*. New York: Routledge.

Atran, S., & Norenzayan, A. (2004). Religion's evolutionary landscape: Counterintuition, commitment, compassion, communion. *Behavioral and Brain Sciences*, *27*, 713–770.

Baerveldt, C., & Verheggen, T. (2012). *Enactivism*. In J. Valsiner (Ed.), *The Oxford handbook of culture and psychology* (pp. 165–190). New York: Oxford.

Baerveldt, C., & Voestermans, P. (2005). Cultural psychology and the normative structure of reality. *Theory and Psychology*, *15*, 449–473.

Bakhtin, M. (1981). *The dialogic imagination: Four essays by M. M. Bakhtin* (C. Emerson & M. Holquist, Trans.). Austin, TX: University of Texas Press. (Original Russian Publication 1975)

Bakhtin, M. (1986). *Speech genres and other late essays* (V. McGhee, Trans.). Austin, TX: University of Texas Press. (Original Russian Publication 1970–1979)

Bakhtin, M. (1990). Author and hero in aesthetic activity (V. Liapunov & K. Brostrom, Trans.). In M. Holquist & V. Liapunov (Eds.), *Art and answerability: Early philosophical essays*. Austin, TX: University of Texas Press. (Original Russian Publication 1979)

Bakhtin, M. (1993). *Toward a Philosophy of the Act* (V. Liapunov, Trans.). Austin, TX: University of Texas Press. (Original Russian Publication 1986)

Ball, P. (2012). Is rationality the enemy of religion? *Nature*. doi:10.1038/nature.2012.10539. Retrieved from www.nature.com.

Barrett, H., & Kurzban, R. (2006). Modularity in cognition: Framing the debate. *Psychological Review*, *113*(3), 628–647.

Barrett, J. (2004). *Why would anyone believe in god?* Plymouth: AltaMira.

Barrett, J. (2007). Cognitive science of religion: What is it and why is it? *Religion Compass*, *1*(6), 768–786.

Barrett, J. (2008). Coding and quantifying counterintuitiveness in religious concepts: Theoretical and methodological reflections. *Method and Theory in the Study of Religion*, *20*, 308–328.

Barrett, J. (2012). *Born believers: The science of children's religious belief*. Toronto: Free Press.

Barrett, J., Burdett, E. R., & Porter, T. (2009). Counterintuitiveness in folk tales: Finding the cognitive optimum. *Journal of Cognition and Culture*, *9*, 271–287.

Barrett, J., & Richert, R. (2003). Anthropomorphism or preparedness? Exploring children's god concepts. *Review of Religious Research*, *44*(3), 300–312.

Barrett, J., Richert, R., & Driesenga, A. (2001). God's beliefs versus mother's: The development of non-human agent concepts. *Child Development*, *72*(1), 50–65.

Belzin, J. (1997). The historicultural approach in the psychology of religion: Perspectives for interdisciplinary research. *Journal for the Scientific Study of Religion*, *36*(3), 358–371.

Belzin, J. (1999). Religion as embodiment: Cultural-psychological concepts and methods in the study of conversion among "Bevindelijken." *Journal for the Scientific Study of Religion*, *38*(2), 36–253.

Belzin, J. (2005). A way out of crisis? From völkerpsychologie to cultural psychology of religion. *Theory & Psychology*, *15*, 812–838.

Belzin, J. (2010a). Religion and self: Notion from a cultural psychological perspective. *Pastoral Psychology*, *59*, 399–409.

Belzin, J. (2010b). *Towards cultural psychology of religion: Principles, approaches, applications*. New York: Springer.

Benson, C. (1991). *The cultural psychology of self*. London: Routledge.

Bering, J. (2006). The folk psychology of souls. *Behavioral and Brain Sciences*, *29*, 453–498.

Bering, J. (2010). *God instinct: The psychology of souls, destiny, and the meaning of life*. London: Nicholas Brealey.

Bering, J. (2011). *The belief instinct: The psychology of souls, destiny, and the meaning of life*. New York: W. W. Norton & Co.

Bering, J., Blasi, C., & Bjorklund, D. (2005). The development of 'afterlife' beliefs in religiously and secularly schooled children. *British Journal of Developmental Psychology*, *23*, 587–607.

Bhabha, H. (1994). *The location of culture*. New York: Routledge.

Billig, M. (1996). *Arguing and thinking: A rhetorical approach to social psychology* (2nd ed.). New York, NY: Cambridge University Press.

Bishop, R. (2007). *The philosophy of the social sciences*. New York: Bloomsbury Academic.

Boden, M. (2006). *Mind as machine: A history of cognitive science*. New York: Oxford University.

Boyer, P. (2001). *Religion explained*. New York: Basic books.

Boyer, P., & Ramble, C. (2001). Cognitive templates for religious concepts: Cross-cultural evidence for recall of counter-intuitive representations. *Cognitive Science*, *25*, 535–564.

Brown, S. (2012, June 26). Analytical thinking promotes disbelief in experimental psychology. *Huffington Post*. Retrieved from www.huffingtonpost.com

Brown, S., & Stenner, P. (2009). *Psychology without foundations: History, philosophy and psychosocial theory*. Thousand Oaks, CA: Sage.

Bruner, J. (1990). *Acts of meaning*. Cambridge, MA: Harvard University Press.

Buller, D. (2005). *Adapting minds: Evolutionary psychology and the persistent quest for human nature*. Cambridge, MA: MIT Press.

Burkitt, I. (2013). Self and others in the field of perception: The role of micro-dialogue, feeling, and emotion in perception. *Journal of Theoretical and Philosophical Psychology*, *33*(4), 267–279.

Cerbone, D. (2006). *Understanding phenomenology*. Chesham: Acumen.

Chirkov, V. (2016). *Fundamental of research on culture and psychology*. New York: Routledge.

Chirkov, V. I. (2009). Critical psychology of acculturation: What do we study and how do we study it, when we investigate acculturation? *International Journal of Intercultural Relations*, *33*(2), 94–105.

Cohen, E., Lanman, J., & Whitehouse, H. (2008). Common criticisms of the cognitive science of religion – answered. *The Council of Societies for the Study of Religion Bulletin*, *37*(4), 112–115.

Cole, M. (2003). *Cultural psychology: A once and future discipline*. Cambridge, MA: Harvard University Press.

Cole, M. (2006). Culture and cognitive development in phylogenetic, historical, and ontogentic perspective. In W. Damon & R. Lerner (Eds.), *Handbook of child psychology Volume II* (pp. 636–683). Hoboken, NJ: Wiley.

Corcoran, T. (2009). Second nature. *British Journal of Psychology*, *48*(2), 375–388.

Cosmides, L., & Tooby, J. (1992). Cognitive adaptations for social exchange. In J. Barkow, L. Cosmides, & J. Tooby (Eds.), *The adapted mind: Evolutionary psychology and the generation of culture* (pp. 163–228). New York: Oxford University.

Cosmides, L., & Tooby, J. (1994). Origins of domain specificity: The evolution of functional organization. In L. Hirschfels & S. Gellman (Eds.), *Mapping the mind: Domain specificity in cognition and culture* (pp. 85–116). Cambridge: Cambridge University.

Cosmides, L., & Tooby, J. (1997). *Evolutionary psychology: A primer*. Center for Evolutionary Psychology, University of California Santa Barbara. Retrieved from www.cep.ucsb.edu/primer.html

Cresswell, J. (2011). Being faithful to ourselves: Bakhtin and a potential postmodern psychology of self. *Culture & Psychology*, *17*, 462–479.

Cresswell, J. (2012). Including social discourses and experience in research on refugees, race, & ethnicity. *Discourse & Society*, *23*(5), 553–575.

Cresswell, J. (2014). Can psychology & religion get along: Towards a pragmatic cultural psychology of religion that includes meaning and experience. *Theoretical and Philosophical Psychology*, *34*(2), 133–145. doi:10.1037/a0033042

Cresswell, J., & Baerveldt, C. (2011). Bakhtin's realism, embodiment, and the dialogical self: Towards a re-visioning of the dialogical self. *Culture & Psychology*, *17*, 263–277.

Cresswell, J., & Baerveldt, C. (2016). On being good researchers: Virtue, sympathetic co-experience and polyphonic unmasking. In T. Corcoran & J. Cromby (Eds.), *Joint action: Essays in honour of John Shotter* (pp. 102–119). Flourance: Routledge.

Cresswell, J., & Smith, L. (2012). Embodying discourse analysis: Lessons learned about epistemic & ontological psychologies. *Discourse & Society*, *23*(5), 619–625.

Cresswell, J., & Teucher, U. (2011). Embodiment and language: M. M. Bakhtin on ontogenetic development. *New Ideas in Psychology*, *29*, 106–118.

Cresswell, J., Wagoner, B., & Haye, A. (2017). William James reconsidered: Introduction to a special issue commemorating the 125th Anniversary of 'Principles of Psychology'. *New Ideas in Psychology*, *46*, 1–6.

Cresswell, J., Wenner, B., & Howell, M. (2014, November). *New vistas from old theories: Working through the complexities of religious belief through radical empiricism*. Paper Presented at the Society for the Scientific Study of Religion Conference, Indianapolis.

Cromby, J. (2015). *Feeling bodies*. New York: Palgrave Macmillan.

Danziger, K. (1990). *Constructing the subject*. New York: Cambridge University Press.

Dawkins, R. (2006). *The God delusion*. Boston, MA: Houghton Mifflin.

Dawson, M. (2001). *Understanding cognitive science*. Malden, MA: Blackwell.

Dennet, D. (1996). *The intentional stance*. Cambridge, MA: MIT Press.

Dennet, D. (2006). *Breaking the spell: Religion as natural phenomenon*. New York: Viking.

Derksen, M. (2007). Cultivating human nature. *New Ideas in Psychology*, *25*, 189–206.

Derksen, M. (2010). Realism, relativism, and evolutionary psychology. *Theory & Psychology*, *20*, 189–206.

Donald, M. (2001). *A mind so rare: The evolution of human consciousness*. New York: Norton.

Edman, L. (2014). *The faithful brain: Using the science of faith in our faith practice*. Psychology and Spiritual Formation: The 3rd Annual Conference of the Center for Christian Thought, La Mirada, CA.

Edman, L. (2015). Applying the science of faith: The cognitive science of religion and Christian practice. *Journal of Psychology and Christianity*, *34*, 238–249.

Edman, L., Vermeer, J., Chambers, H., & Wiggers, T. (2015). *Mentalizing and prayer: The relationship of theory of mind and knowing a personal god*. The Society for the Scientific Study of Religion 2015 Annual Conference, Newport Beach, CA.

Edwards, D. (1997). *Discourse and cognition*. Thousand Oaks, CA: Sage.

Elman, J., Bates, E., Johnson, M., Karmiloff-Smith, A., Parisi, D., & Plunkett, K. (1998). *Rethinking innateness: A connectionist perspective on development*. Cambridge, MA: MIT Press.

Fogel, A. (1993). *Developing through relationships: Origins of communication, self and culture*. New York: Harvester.

Fowers, B. (2005). *Virtue and psychology: Pursuing excellence in ordinary practices*. Washington, DC: American Psychological association.

Fowers, B. (2015). The promise of a flourishing theoretical psychology. *Journal of Theoretical & Philosophical Psychology*, *35*(3), 145–159.

Friesen, L., & Cresswell, J. (2017). Rethinking priming: Insight from James' notions of habits and instincts. *New Ideas in Psychology*, *46*, 17–25.

Gadamer, H. (2008). *Philosophical hermeneutics*. Los Angeles, CA: University of California.

Gantt, E., Melling, S., & Reber, J. (2012). Mechanisms or metaphors? The emptiness of evolutionary psychological explanations. *Theory and Psychology*, *22*(6), 823–841.

Geertz, C. (1973). *The interpretation of cultures*. New York: Basic Books.

Gergen, K. (1985). The social constructionist movement in modern psychology. *American Psychologist*, *40*(3), 266–275.

Gergen, K. (1994). *Realities and relationships*. Cambridge, MA: Harvard University Press.

Gergen, K. (1999a). Agency: Social construction and relational action. *Theory and Psychology*, *9*(1), 113–115.

Gergen, K. (1999b). *An invitation to social construction*. Thousand Oaks, CA: Sage.

Gervais, W., & Norenzayan, A. (2012). Analytic thinking promotes disbelief. *Science*, *336*, 493–496.

Greenspan, S., & Shanker, S. (2006). *The first idea: How symbols, language, and intelligence evolved from our primate ancestors to modern humans*. Cambridge, MA: De Capo.

Greenspan, S., & Shanker, S. (2007). The developmental pathways leading to pattern recognition, joint attention, language and cognition. *New Ideas in Psychology*, *25*, 128–142.

Gregory, J., & Barrett, J. (2009). Epistemology and counterintuitiveness: Role and relationship in epidemiology of cultural representations. *Journal of Cognition & Culture*, *9*, 289–314.

Guthrie, S. (1993). *Faces in the clouds: A new theory of religion.* New York: Oxford University.

Guthrie, S. (2002). Animal animism: Evolutionary roots of religious cognition. In I. Pyysiäinen & V. Anttonen (Eds.), *Current approaches in the cognitive science of religion* (pp. 38–67). New York: Continuum.

Hall, D., Meador, K., & Koenig, H. (2008). Measuring religiousness in health research: Review and critique. *Journal of Religious Health, 47*, 134–163.

Halonen, J., Landrum, E., Buskist, W., BMcCarthy, M., Dunn, D., Rudmann, J., Freeman, J, Stoloff, M., Hill, W., Enns, C., & Kaslow, N. (2013). *APA guidelines for the undergraduate psychology major – Version 2.0.* On-line document prepared by the American Psychological Association Board of Educational Affairs Task Force on Psychology Major Competencies. Retrieved fromwww.apa.org/ed/precollege/about/psymajor-guidelines.pdf

Hampson, P. (2005). Cultural psychology and theology: Partners in dialogue. *Theology and Science, 3*(4), 259–274.

Harré, R. (2002). *Cognitive science*. Thousand Oaks, CA: Sage.

Hermans, H., & Kempen, H. (1993). *The dialogical self.* Toronto: Academic Press.

Hirschfeld, L., & Gellman, S. (1994). Towards a topography of mind: An introduction to domain specificity. In L. Hirschfeld & S. Gellman (Eds.), *Mapping the mind: Domain specificity in cognition and culture* (pp. 3–36). Cambridge: Cambridge University.

Hofstede, G. (1980). *Culture's consequences: International differences in work-related values*. Thousand Oaks, CA: Sage.

Hsu, H., & Fogel, A. (2002). Stability and transitions in mother-infant face-to-face communication during the first 6 months: A microhistorical approach. *Development Psychology, 39*(6), 1061–1082.

James, W. (1956). *Will to believe*. New York: Dover Publications (original publication 1897)

James, W. (1981). *Pragmatism*. Indianapolis, IN: Hackett (original publication 1907)

James, W. (1982). *The varieties of religious experience*. New York: Penguin (original publication 1902)

James, W. (1996). *A pluralistic universe*. Lincoln: University of Nebraska Press (original publication 1909)

James, W. (1996). *Essays in radical empiricism*. Lincoln: University of Nebraska Press (original publication 1912)

James, W. (2011). *The meaning of truth*. New York: Barnes & Noble Digital Library (original publication 1909)

Kapuscinski, A., & Masters, K. (2010). The current status of measures of spirituality: A critical review of scale development. *Psychology of Religion & Spirituality, 2*(4), 191–205.

Karasavvidis, I. (2002). Distributed cognition and educational practice. *Journal of Interactive Learning Research, 13*, 11–29.

Kashima, Y. (2000). Conception of culture and person for psychology. *Journal of Cross-Cultural Psychology, 31*(1), 14–32.

Kashima, Y. (2016). Culture and psychology in the 21st century: Conceptions of culture and person for psychology revisited. *Journal of Cross-Cultural Psychology, 47*(1), 4–20.

Keesing, R. (1974). Theories of culture. *Annual Review of Anthropology, 3*, 73–97.

Keesing, R. (1990). Theories of culture revisited. *Canberra Anthropology, 13*(2), 46–60.

Keleman, D. (2004). Are children "Intuitive theists?" Reasoning about purpose and design in nature. *Psychological Science, 15*(5), 296–301.

Keleman, D., & Rosset, E. (2009). The human function compunction: Teleological explanation in adults. *Cognition, 111*, 138–143.

Kitayama, S., & Cohen, D. (Eds.). (2007). *Handbook of cultural psychology*. New York: Guilford Press.

Krakovsky, M. (2012, April 26). Losing your religion: Analytic thinking can undermine belief. *Scientific American*. Retrieved from www.scientificamerican.com

Lewis, M. D. (2000). The promise of dynamic systems approaches for an integrated account of human development. *Child Development*, 71(1), 36–43.

Maturana, H., & Varela, F. (1980). *Autopoiesis and cognition*. Hingham, MA: Reidel Publishing.

McCauley, R. N. (2011). *Why religion is natural and science is not*. New York: Oxford University.

McCauley, R. N., & Lawson, E. (2002). *Bringing ritual to mind: Psychological foundations of cultural forms*. Cambridge: Cambridge University Press.

McConnel, K., & Edman, L. (2013). *Theological correctness, practiced naturalism, and the intrusion of intuitive concepts*. The Society for the Scientific Study of Religion Annual Conference, Boston, MA.

McLean, M., Cresswell, J., & Ashley, C. (2016). Psychologists finding religion: Enhancing cognitive science of religion with cultural psychology. *Culture & Psychology, 22*(1), 44–64.

Meezenbroek, E., Garssen, B., van den Berg, M., van Dierendonck, D., Visser, A., & Schaufeli, W. (2012). Measuring spirituality as a universal experience: A review of spirituality questionnaires. *Journal of Religion and Health, 31*, 336–354.

Miller, P. (1996). Instantiating culture through discourse practices: Some personal reflections on socialization and how to study it. In R. Jessor, A. Colby, & R. Shweder (Eds.), *Ethnography and human development* (pp. 183–204). Chicago: University of Chicago Press.

Murray, M., & Goldberg, A. (2009). Evolutionary accounts of religion: Explaining and explaining away. In J. Schloss & M. Murray (Eds.), *The believing primate: Scientific, philosophical and theological reflection on the origins of religion* (pp. 100–117). Oxford: Oxford University Press.

Neisser, U. (1994). Multiple systems: A new approach to cognitive theory. *European Journal of Cognitive Psychology, 6*(3), 225–241.

Nietzsche, F. (1964). *Nietzsche: An anthology of his works*. New York: Washington Square Press.

Noë, A. (2009). *Out of our heads: Why you are not your brain and other lessons from the biology of consciousness*. New York: Hill & Wang.

Nussbaum, M. (2016). *Not for profit: Why democracy needs the humanities*. Princeton, NJ: Princeton University Press.

Oppenheim, D., Nir, A., Warren, S., & Emde, R. N. (1997). Emotion regulation in mother – child narrative co-construction: Associations with children's narratives and adaptation. *Developmental Psychology, 33*(2), 284–294.

Paloutzian, R. (1996). *Invitation to the psychology of religion* (2nd ed.). Toronto: Allyn & Bacon.

Pappas, S. (2012, April 26). Analytic thinking can promote atheism. *LiveScience*. Retrieved from http://www.livescience.com

Peters, B. (2013). Evolutionary psychology: Neglecting neurobiology in defining the mind. *Theory and Psychology, 23*(3), 305–322.

Polkinghorne, D. E. (1983). *Methodology for the human sciences: Systems of inquiry*. Albany: State University of New York Press.

Potter, J., & Wetherell, M. (1987). *Discourse and social psychology: Beyond attitudes and behaviour*. Newbury Park: Sage.

Prinz, J. (2009). Is consciousness embodied? In P. Robbins & M. Aydede (Eds.), *The Cambridge handbook of situated cognition* (pp. 419–537). New York: Cambridge University Press.

Proulx, J. (2004). The enactivist theory of cognition and behaviorism: An account of the processes of individual sense making. *Proceedings of the 2004 Complexity Science and Research Conference* (pp. 115–120).

Pyysiäinen, I. (2002). Introduction: Cognition and culture in the construction of religion. In I. Pyysiäinen & V. Anttonen (Eds.), *Current approaches in the cognitive science of religion* (pp. 1–13). New York: Continuum.

Ratner, C. (2006). *Cultural psychology: A perspective on psychological functioning and social reform*. Mahwah, NJ: Psychology Press.

Reber, J. (2006). Secular psychology: What's the problem? *Journal of Psychology and Theology*, *34*(3), 193–204.

Richardson, F. (2012). On psychology and virtue ethics. *Journal of Theoretical and Philosophical Psychology*, *32*(1), 24–34.

Richardson, R. (2007). *Evolutionary psychology as maladapted psychology*. Cambridge, MA: MIT Press.

Richerson, P., & Newson, L. (2009). Is religion adaptive? Yes, no, neutral. But mostly we don't know. In J. Schloss & M. Murray (Eds.), *The believing primate: Scientific, philosophical and theological reflection on the origins of religion* (pp. 100–117). Oxford: Oxford University Press.

Robbins, P., & Aydede, M. (2009). A short primer on situated cognition. In P. Robbins & M. Aydede (Eds.), *The Cambridge handbook of situated cognition* (pp. 3–10). New York: Cambridge University Press.

Rogoff, B. (2005). *The cultural nature of human development*. Oxford: Oxford University Press.

Rohner, R. P. (1984). Toward a conception of culture for cross-cultural psychology. *Journal of Cross-Cultural Psychology*, *15*(2), 111–138.

Rosaldo, M. (1984). Toward an anthropology of self and feeling. In R. Shweder & R. LeVine (Eds.), *Culture theory: Essays on mind, self, and emotion*. New York: Cambridge.

Said, E. (1996). *Representations of the intellectual*. New York: Vintage Books.

Saxe, G. (1996).Studying cognitive development in sociocultural context: The development of a practice-based approach. In R. Jessor, A. Colby, & R. Shweder (Eds.), *Ethnography and human development* (pp. 275–304). Chicago: University of Chicago Press.

Segall, M., Lonner, W., & Berry, J. (1998). Cross-cultural psychology as a scholarly discipline: On the flowering of culture in behavioural research. *American Psychology*, *53*(10), 1101–1110.

Shannon, C. E., & Weaver, W. (1949). *The mathematical theory of communication*. Urbana, IL: University of Illinois Press

Shapiro, L. (2011). *Embodied cognition*. New York: Routledge.

Shor, I. (1987). *Critical teaching and everyday life*. Chicago: University of Chicago Press.

Shotter, J., & Billig, M. (1998). A Bakhtinian psychology: From out of the heads of individuals and into the dialogues between them. In M. Bell & M. Gardiner (Eds.), *Bakhtin and the human sciences: No last words* (pp. 13–29). Thousand Oaks, CA: Sage.

Shweder, R. (1991). *Thinking through cultures: Expeditions in cultural psychology*. Cambridge, MA: Harvard University Press.

Simon, H. (1998). What is an 'explanation' of behavior? In P. Thagard (Ed.), *Mind readings: Introductory selections on cognitive science* (pp. 1–28). Cambridge, MA: MIT Press.

Slife, B., & Reber, J. (2009). Is there a pervasive implicit bias against theism in psychology? *Journal of Theoretical and Philosophical Psychology*, *29*(2), 63–79.

Slife, B., & Reber, J. (2012). Conceptualizing religious practices in psychological research: Problems and prospects. *Pastoral Psychology*, *61*, 735–746.

Slife, B., & Williams, R. (1995). *What's behind the research?* Thousand Oaks, CA: Sage.

Slone, D. (2004). *Theological correctness: Why religious people believe what they shouldn't*. New York: Oxford University Press.

Smith, C. (2010). *What is a person: Rethinking humanity*. Chicago: University of Chicago.

Smith, P., & Bond, M. (1999). Culture: The neglected concept. In P. Smith & M. Bond (Eds.), *Social psychology across cultures* (pp. 38–68). Harlow: Prentice Hall.

Spelke, E. S., Breinlinger, K., Macomber, J., & Jacobson, K. (1992). Origins of Knowledge. *Psychological Review*, 99(4), 605–632.

Spelke, E. S., & Kinzler, K. (2007). Core knowledge. *Developmental Science*, *10*(1), 89–96.

Sperber, D. (1994). The modularity of thought and the epidemiology of representations. In L. Hirschfeld & S. Gellman (Eds.), *Mapping the mind: Domain specificity in cognition and culture* (pp. 39–67). Cambridge: Cambridge University.

Sperber, D. (1996). *Explaining culture: A naturalistic approach*. Malden, MA: Blackwell.

Sperber, D., & Hirschfeld, L. (2004). The cognitive foundations of stability and diversity. *Trends in Cognitive Science*, *8*(1), 40–46.

Stam, H. (2015a). The historical boundedness of psychological knowledge and the ethics of shared understandings. *Journal of Theoretical & Philosophical Psychology*, *35*(2), 117–127.

Stam, H. (2015b). Indeterminate functionalism and possessive individualism. In J. Cresswell, A. Haye, A. Larraín, M. Morgan, & G. Sullivan (Eds.), *Dialogue and debate in the making of theoretical psychology* (pp. 385–403). Concord, ON: Captus.

Sugarman, J. (2015). Neoliberalism and psychological ethics. *Journal of Theoretical & Philosophical Psychology*, *35*(2), 105–116.

Tallis, R. (2011). *Aping mankind: Nueromania, darwinitis, and the misrepresentation of humanity*. Durham: Acumen.

Taylor, C. (1989). *Sources of the self: The making of the modern identity*. Cambridge, MA: Harvard University Press.

Taylor, C. (1999). *Human agency and language: Philosophical papers volume 1*. Cambridge: Cambridge University Press.

ten Have, P. (2002). *Doing conversation analysis: A practical guide*. Thousand Oaks, CA: Sage.

Teo, T. (2008). From speculation to epistemological violence in psychology: A critical-hermeneutic reconstruction. *Theory & Psychology*, *18*(1), 47–67.

Teo, T. (2015). Are psychological "ethics codes" morally oblique? *Journal of Theoretical & Philosophical Psychology*, *35*(2), 78–89.

Thagard, P. (2005). *Mind: Introduction to cognitive science*. Cambridge, MA: MIT Press.

Thelen, E. (2005). Dynamic systems theory and the complexity of change. *Psychoanalytic Dialogues*, *15*(2), 255–283.

Thelen, E., & Smith, L. (1996). *A dynamic systems approach to the development of cognition and action*. Cambridge, MA: MIT Press.

Thompson, E. (2007). *Mind in life: Biology, phenomenology, and the sciences of mind.* Cambridge, MA: Harvard University.
Tooby, J., & Cosmides, L. (1992). The psychological foundations of culture. In J. Barkow, L. Cosmides, & J. Tooby (Eds.), *The adapted mind: Evolutionary psychology and the generation of culture* (pp. 19–136). New York: Oxford University.
Underwood, L. (2011). The daily spiritual experience scale: Overview and results. *Religions, 2*, 29–50.
Underwood, L., & Teresi, J. (2002). The daily spiritual experience scale: Development, theoretical description, reliability, exploratory factor analysis, and preliminary construct validity using health-related data. *Annals of Behavioral Medicine, 24*(1), 22–33.
Valinser, J. (2012). *The Oxford handbook of cultural psychology*. New York: Oxford University.
van Orden, G. C., Holden, J. G., & Turvey, M. T. (2003). Self-organization of cognitive performance. *Journal of Experimental Psychology: General, 132*(3), 331–350.
Varela, F., Thompson, E., & Rosch, E. (1992). *The embodied mind: Cognitive science and human experience*. Cambridge, MA: The MIT Press.
Vattimo, G. (2002). *After Christianity*. New York: Columbia University.
Vattimo, G. (2011). *A farewell to truth*. New York: Columbia University.
Walsh, R. (2015). Bending the arc of North American psychologists' moral universe towards communicative ethics and social justice. *Journal of Theoretical & Philosophical Psychology, 35*(2), 90–102.
Walsh, R., & Gokani, R. (2014). The personal and political economy of psychologists' desires for social justice. *Journal of Theoretical & Philosophical Psychology, 34*(1), 41–55.
Wertsch, J. (1985). *Vygotsky and the social formation of mind*. Cambridge, MA: Harvard University Press.
Wertsch, J. (1991). *Voices of the mind: A sociocultural approach to mediated action*. London: Harvester Wheatsheaf.
Whitehouse, H. (2004). *Modes of religiosity: A cognitive theory of religious transmission*. New York: AltaMira Press.
Xygalatas, D. (2012). *The burning saints: Cognition and culture in fire-walking rituals of the Anastenaria*. Sheffield: Equinox.
Yonker, J., Edman, L., Cresswell, J., & Barrett, J. (2016). Primed analytic thought and religiosity: The importance of individual characteristics. *Journal of Religion and Spirituality*. On-line pre-print, doi:10.1037/rel0000095.
Žižec, S. (2008). *The plague of fantasies*. New York: Verso.

Index

For Product Safety Concerns and Information please contact our EU representative GPSR@taylorandfrancis.com
Taylor & Francis Verlag GmbH, Kaufingerstraße 24, 80331 München, Germany

www.ingramcontent.com/pod-product-compliance
Ingram Content Group UK Ltd.
Pitfield, Milton Keynes, MK11 3LW, UK
UKHW022122190926
475702UK00013B/1594

9780367363383